RETURN TO THE MOON

Edited by Rick Tumlinson
with Erin Medlicott

An Apogee Books Publication

We acknowledge the financial support of the Government of Canada through the Book Publishing Industry Development Program for our publishing activities.

Published by Apogee Books, Box 62034, Burlington, Ontario, Canada, L7R 4K2.
http://www.apogeebooks.com

Printed and bound in Canada

Return To The Moon
Edited by Rick Tumlinson with Erin Medlicott

ISSN-1496-6921
ISBN 1-894959-32-9 (1st Edition)

CONTENTS

Editor's Note - This book, and many of the essays contained herein, are admittedly written from a US centric point of view. In my case and that of most of the authors, it is not because we choose to exclude other nations and peoples from this great adventure, nor is it out of ego or hubris. Rather, it is because it is where we live, the government and systems we can influence, and the culture we can most greatly affect. I urge you not to take offence, as this is a cause for all humanity. Listen to the meaning behind the words, see the bigger picture and go out into your own culture or nation and push this concept in ways that work for you. We will meet at my place on the Moon, where all will be welcome, and tip a glass to that green orb we all call home.

The Moon
by Rick Tumlinson

It was once said that if God had wanted humanity to go into space we would have been given a Moon. And we were. It draws our eyes upwards at night, and unlike the distant glimmer of the countless stars that fill the darkness, it is close, it feels like it is right there, with mountains and terrain just like our own world that you can almost see with your own eyes. The Moon is an undeniable reminder that there is more beyond the Earth, that there are other worlds out there - that we are not alone. So strong is its pull that it eventually drew us outwards as the first place we would go beyond this Earth.

So we went. Some thirty years ago, after a decade of the most ambitious and intense effort in human history, the people of Earth reached out and touched our silvery neighbor. As a teenager I watched as over the course of several years small groups of our explorers bounced across its cold grey surface, snapping pictures, shooting film, taking samples… The effort to get there had been thrilling, dangerous, the stuff of myths and legends the type of activity that creates heroes and inspires generations. Yet when they finally arrived, what they saw….the soft grey surface rolling off into the pitch black horizon of space, and what they found, gathered in now dusty sample containers wasting away on the shelves of some space facility, was not enough to hold the interest of the world. The Moon was a black and white enigma in the dawning age of color TV. So after leaving a few flags and footprints on its dusty surface, in 1972 the people of Earth turned away…they had other things to do it seemed…and so did I - after all it was the '70s.

And so the thirty years passed, and human space activity fell off the cultural map and into the oblivion of timidity and broken promises, punctuated by tragedy and an occasionally interesting stunt. Then, the winter of 2004 I received a call at around 6:30 in the morning pacific time. Half asleep in the darkness I heard a far too young, too wide awake and too bouncy sounding voice on the other end say something about his name being Chip, and that he was calling from the White House. He said the president would like me to attend his remarks the next day at a private event where he would announce we were going back to the Moon. Frankly, I had basically tuned out around the words White House, mumbled "uh huh…yeah…" hung up the phone and rolled over to go back to sleep.

A few minutes later, as the weirdness of the semi-conversation began to sink in, it occurred to me that this might not be a prank call from one of my east coast friends, who are forever giving me grief about being a late sleeper (this compounded by my being on the west coast). I groggily called one of my pals in DC and asked if he knew anything about such an announcement, which had been rumoured for months. He said yes, it was happening, and made the intelligent suggestion I check my caller ID…which, being still half asleep, had not occurred to me.

I staggered into my office, looked at the number, and well, to shorten a much longer story, 24 hours later was walking into a lecture hall at NASA headquarters with a bunch of NASA, government and aerospace leaders (several of whom, knowing me from pictures in space journals and on their corporate dart boards had puzzled looks on their faces). Worse, I found out as I was ushered in that I was one of 20 guests invited by the White House itself, and only one of a very few from space organizations.

After a suitably forgettable speech by the then NASA administrator, President Bush entered the room. Three rows from the front I watched as he laid out what has since been labeled his Vision for Space Exploration, a plan to send people back to the Moon and then use that world as a test bed, training camp and launching pad for the exploration of Mars and beyond. As I later testified to the Senate, I looked closely at his eyes and listened closely to his words as he spoke. I did so knowing he was a top-level politician, with all the skills of persuasion and misdirection that accompany such a job. Yet, from all I could tell, what he said seemed to be coming from his heart. He meant it. We were going back to the Moon! In this speech he was giving our space agency its marching orders, aiming them at a destination and putting them back into the exploration game. This was the first time they had been given such a clear mandate since John Kennedy kicked off Apollo, and it was an exciting moment-even given the background of political, economic and bureaucratic realities which would soon try to swallow and warp his words.

The timing of this speech was both bad and good. It was bad because it was an election year and some took it as a cheap way to gain votes, which I am convinced it was not (one would expect his very savvy advisors to know that it would draw fire and ridicule). The timing was also bad as the economic news of the day was full of stories about a huge deficit-mainly blamed on the same president making the speech. But it was a perfect time as well. In fact, such a speech had to made within a limited window if it was to be effective in causing maximum change. The Columbia disaster had nearly wiped out faith in NASA, showing how little had been learned since Challenger. As former administrator Thomas Paine once told my compatriots and I, the inevitable consequence of "flying shuttles in circles until another one falls out of the sky..." had occurred. The NASA "can-do" brand had been seriously damaged - again. Its impenetrable walls of institutional incompetence and bureaucratic cover-your-ass turf protection were vulnerable and crumbling under the spotlight of the accident and its follow-up investigations. Long badly led and guided by the clueless and self-promoting, the agency's good people needed hope and a new direction that would force those without energy and vision out the door and cultivate those with the long missing "right stuff" spirit that had been its brand before the accident. It was time for change, time for leadership and real direction that would put the agency back into its Lewis and Clark role. And here it was. A set of tough but clear goals, and little money to do it. As the Chinese might say, it was an opportunity to be creative.

Although the president had little or no concept of it at the time, and may still not understand, his timing was important for another reason, one which in the long view may well eclipse that of revitalizing and focusing our government space effort.

Space Settlement and the Frontier Movement

You see, since the mid-1970s a small but growing movement has been planting the seeds of a revolution in space. A revolution based on people opening space for themselves, by themselves and with their own money, sweat and ingenuity. I call this the space frontier movement, and since the late 1990s it has begun to show itself as a force with ever-growing momentum. They are led by a NewSpace industry of visionary people who are rolling up their sleeves and making it happen. They are inspired by the same dream and the same core goals, expressed in the writings and political actions of their fellow activists, and supported by investors, workers and volunteers who believe. This is a true revolution. It is not just about the Moon, or Mars or any single destination. It is a revolution in the way we not only see humanity, but directs how we should act in relation to our world, our future and indeed our destiny as a species.

From my perspective, the core concepts of this movement are simple. Earth is a cradle of life in a potentially hostile universe. Humanity is the caretaker of that life. It is our job to not only save this world but to do so using the resources of the space around us, and to take life to worlds now dead. To expand the human species and life forms of Earth beyond the limiting confines of this tiny world and out into the universe.

To those in the movement this is not far-out rhetorical and poetic fluff. It is what we are about. It is who we are. It is what we are doing. And now, in the last few years, we have arrived.

The president was not the only one in government who didn't yet fully understand what was really happening in space beyond the confines of government programs. Many others who should know were just waking up to the fact that we are in a new ballgame. For example, a month after the president's speech, I was asked to testify in front of John McCain and the Senate Science Committee. I began my testimony by stating that before the end of 2004 the next American (since Columbia) would fly in space, that they would not be a government employee, and they would not be flying on a government rocket. The Senator, having been lulled into near unconsciousness by previous testimony from NASA and others, sat bolt upright and asked me to explain...I told him to wait and see.

And see he did. Within 6 months of my words, thanks to the work of my co-conspirators in the XPrize (Peter Diamandis and Greg Maryniak were the leaders, I was a founding Trustee), Burt Rutan had sent the tiny white bird of SpaceShip One up into the edge of space. His flights, and the resulting publicity, combined with the flight a few years earlier by Dennis Tito to the international space station that I and others, had arranged, finally came together in the dawning of a new idea-space was no longer the exclusive domain of government employees and their programs.

With billionaires and business people pouring hundreds of millions into private space travel, we are only a few years from seeing the development of a major and historic breakout into space. Already we hear of commercial passenger flights around the Moon, and the first space hotels are being built right now. Within five or six years the next big turning point will occur. At that time the first commercial orbiting space facilities will be operational-and serviced by commercial spacecraft.

Why is that important? Because at that moment the government program is no longer needed if our goal is the opening of space to humanity. The "budding" of this new branch will have been successful.

And why is all of this relevant to a book on Returning to the Moon? Several reasons.

The president, by laying out his Vision for Space Exploration has set the government space program on a course, which, if pursued correctly, will result in the establishment of the first human communities beyond Earth orbit. This goal, although not clear to all involved in this great task, is the implied and correct result of this program-and the investment to be made by taxpayers-who will not and should not be asked to support another flags and footprints programs such as Apollo-be it on one world or two.

And at least on the government side of the equation, that is the million-dollar (or billion dollar in this case) question. Faced with a Presidential challenge to take Americans to the Moon and Mars on a budget, our space agency has predictably chosen to answer what is really an economic or business management question with an engineering solution. NASA could have taken the long view, wherein the leverage of national exploration dollars are used to catalyze an industrial and transportation infrastructure that would both dramatically lower the cost of getting into space, and also allow us to stay and expand our presence over time. Instead, rather than seize the opportunity to be truly creative that is offered by trying to develop a permanent human presence on the Moon and Mars within a tight budget, NASA has combined its worse central design bureau instincts with short-sighted timidity in its goals for its plans when it arrives (if ever) at its destinations.

So here we go again. When it comes to NASA's rocket plans for example, not only is there no budget for these constituency feeding dead ends, they are based on an expendable architecture that seems to be designed more to beat arbitrary deadlines than to open the frontier. In other words, just as Columbia showed there was little learned after Challenger, 30 years or so after the last citizens left the Moon, there has been no learning from Apollo.

And that is what I find so frustrating. The agency's plans are doomed to lead inevitability to a repeat of the failure of Apollo. The warning signs are right there, in the form of an ancient flag and dusty grey footprints, or in the sad and angry eyes of our Apollo astronauts and those who gave them flight, in the form of unflown giant white rockets lying at the entrances of our space centers, and yet they are being ignored. We know what happens when a state-dominated, single-point goal takes precedence over a long-term vision. We know what happens when operationally unsupportable technologies win out over economic sustainability. And we know what happens when our government builds one great big door to space, instead of helping Americans blaze a thousand trails to the frontier. We fail. Except this time, if the program doesn't collapse within the next few years of its own unsupportable financial weight, we will have two sets of flags and footprints to show disbelieving future generations. One in shades of grey, and one in shades of red.

Do we go to look or do we go to live? That is the question. Do we build up the systems and architectures which allow us to live beyond the Earth permanently or do we throw together a short-term constituency pleasing set of engineering solutions that are so costly and risky to use that they, like all those out of the same shops before them end up in the trash can of engineering history?

The Vision for Space Exploration as laid out by the president, is simply too expensive for the agency to carry out if it goes down its traditional path, with its traditional partners. They cannot do it. It will fail. Period. And yet, that is exactly what they are doing.

Meanwhile, the NewSpace industry is trying to take flight. Several wealthy entrepreneurs are pouring hundreds of millions into getting it off the ground, but they need help. As in any other industry or commercial enterprise, they need customers and investors. They need expensive research and development support, and they need a level playing field to compete on a global scale with other entities that are often funded by their own governments and supported to a much greater extent than any support offered here at home.

I firmly believe the NewSpace industry can achieve its take-off point of independent access and habitation on its own, but it can do so much more quickly, and be much less vulnerable, if it also has the government as a partner and customer. And speed I believe is important, as so much of what we may find and develop in space will apply to saving life on Earth at a critical time (as you will see in some of the following essays.)

The irony (and there are many in this field) is that a well tailored national and international approach to space, based on frontier principles, using the NewSpace industry and designed to result in permanence, also gives you the lowest long term costs, highest safety margins, feeds all possible constituencies and not only does so on a budget, but a national (and global) profit!

But this requires a significant change in the culture and approach to space activities as currently practiced by our space agencies, particularly here in the US. In fact, it could be said that the biggest challenge NASA faces is not in space, but within its own hallways. The agency must adopt a totally new philosophy of how to succeed in this goal, or it will fail. We may be able to put people on the Moon again, we may even be able to land them on Mars, but we will not be able to keep them there-at least until the private sector gets there on its own-much later than if we do it together.

The Frontier Model

The Frontier model is a helpful tool in understanding the relationships of the public and private sectors in space. It isn't perfect and there are exceptions, but it at least helps clarify who should do what, and where they should place their effort. In this model the private sector and the government develop a partnership, with each focused on its appropriate and traditional role in a free enterprise society.

In the Frontier model it is helpful to visualize the Earth as the center of a bubble of human activity moving outwards into space, the edge of this bubble being the farthest that humans have ventured-in this case the Moon. Within the bubble is the Near Frontier, long since explored by humans and robotic craft, where the private sector is more and more active, commercial activities are underway and private investments are more and more feasible. Beyond the edge of the bubble lies the Far Frontier, where business plans don't yet make sense, infrastructure is non-existent and travel times and mission costs preclude most private concerns from operating. Science and prestige are the drivers, funded by pooling resources in the form of taxes or donations to institutions. These zones represent a continuum, but the Moon is an obvious transition zone, as we have been there, and there are potential areas of economic activity that can arguably be developed.

With the VSE as its driver and the establishment of a human outpost on the Moon as its first step, the US government has the chance to invest tax dollars in ways that will result in not only lowering the cost of explorations of Far Frontier destinations such as Mars, but will also catalyze an industrial base and spur the development of an economic infrastructure leading to the eventual self-sustaining human communities. In the end the investment of funds by the American people will be more than returned in the form of new concepts, products, resources, inspiration and yes, science-for when one is actually living at a place, it is much easier to study.

It is hard for many in the traditional aerospace and science community to trust in this concept, as they do not see an existing base of companies that are engaged in and profiting from space activities. They feel the non-traditional (or New Space) firms are looking for handouts, threaten existing gravy trains, and cannot be trusted with the core work of the mission they feel they have been given by the White House. In one of the most often heard, yet humorously ironic, statements they could make, NASA's knights of the status quo argue that we cannot trust the commercial New Space industry's big promises. They say they don't believe these firms can deliver. Indeed, some do boast a bit too loudly, and many of those wanting to service NASA's needs have not been tested. However, the NASACorp Space Industrial Design Bureau has a long history of promises, it definitely has been tested, and time after time it has failed to deliver.

Let's look at just a few of these: ISS was promised at 6 years to completion for 8 billion dollars. We are now at 17 + years, when finished (if ever) it may well cost $100 billion plus-and is still undelivered. Worse, the day after it is delivered we are tossing it overboard or giving it away to our partners…as NASA cannot afford to run it as it does now while going to the Moon and Mars. (Yet another irony, as the idea of a space station as laid out by Von Braun was as place to assemble and launch missions to other worlds.) Or what about the space shuttle, which was sold on the promise of 50 + flights a year. What was delivered achieves sometime 3 flights a year (when operational and not shut down for blowing up.) Or more recently, we were promised the return to flight of a much safer and robust shuttle fleet. After 2 ½ years and over 2 billion spent not fixing the problem that caused its shut down…what was delivered was…(…never mind, it makes me too crazy to think about it…).

So we have a tested and consistently failed establishment telling us that we should be wary of an untested yet very promising and successfully growing new industry. An industry that isn't even asking for the blind, pay for effort rather than results funding which has characterized the past, but just to be able to fairly compete for jobs and only paid for results. Results. They don't get paid unless they do the job. Results. If they can't prove they can fly astronauts and payloads to space and back, they don't get the contracts. What could be simpler? And what could guarantee an affordable, broadly based and growing space infrastructure better than commercial firms with proven products serving both government and private markets? And why can't the government step up to this new approach rather than relying on those that just never work? Perhaps it is easier to fail using systems you know, than to possibly succeed with those you don't...

For their part, the New Space firms see the government and its select contractors as bloated, inefficient, and unable to deliver economically sustainable space programs and projects on time and within a budget. They and the non-traditional companies who have tried to play in the space game such as construction, mining, and medical research firms have been cheated, blindsided and abused so many times they don't trust the government at all. By shifting to a frontier orientation our government can eliminate this stand off. Instead of competing with each other, both sectors can enhance the other's abilities to deliver on its traditional role in our culture.

Just as here in America the mail was given to the early airlines, or land was granted to those laying the railroads, governments can support the growth of space industries. One of the best ways of doing this, while not overly stretching this new industry beyond its reach is to hand off the management, support and transportation to and from the International Space Station. It is there, in LEO that NewSpace can prove itself, while relieving NASA of its now burdensome operational obligations. Not only can the technologies and operational sides of routine space habitation be developed there, so too can the political, managerial and social structures and systems we will need on the Moon and Mars.

Another solution is for NASA to build into its plans, from the beginning, that as they move on to their own destinations, such as Mars, the infrastructure and resource prospecting, utilization and development work will be developed with the private sector with the goal of a hand-off to them as the government explorers move on. Rather than designing the habs etc., NASA should stay Lewis and Clark-like and focus on such things as scouting expeditions, and establishing an early base camp that is designed to be expandable. Then put out a call to the NewSpace and non-traditional-space community for facilities that are low cost, robust, low maintenance and modular or expandable on a larger scale. NASA and other agencies (American and international) could then sign ten or fifteen year leases, indicating (at least in the case of NASA) they are not planning on squatting down on the Moon but are moving on.

There are a lot of small firms and even a few giant non-space companies out there who want to get involved, from providing the rides, to building and supporting the facilities. Firms who believe they can leverage the effort to grow profitable enterprises-the ultimate mark of success on the frontier. But there is a lack of knowledge regarding how to transition to this model, as well as great mistrust on both sides of the public/private line. Again, the real private sector MUST be involved early on, not as a show, not as an after thought. If the first Moon base is to turn into a settlement or community, it has to be designed to do so from day one.

And therein lies the big question. Does the President's proposal lead to the real opening of the frontier-by which I mean the expansion of the human domain beyond the Earth? Not outposts, not stations, not laboratories, but economically viable and growing communities of human beings that can eventually become new branches of our civilization. For if that is not the end point of this exercise, then, as some in our science community have said repeatedly, we should send robots instead.

Governments do not open frontiers-people do-with the assistance of their governments, and sometimes in spite of those same governments. If this endeavor is to be led by, and for, the government, and simply placing flags on other worlds is its goal, it will fail. Just as the Vikings failed in North America, and we failed the first time we went to the Moon. Our government must learn that this time we go together. It is too important to fail.

The Moon, Mars and the asteroids that float between the worlds of our solar system represent the future of humanity. It would be pure ignorant hubris to declare that we should not expand our species and the domain of life beyond this Earth, much like the declarations of a serf in medieval Europe proclaiming that the world ends just beyond the boundaries of their own village.

Similarly, there is the short-term thinking that leads to the conclusion that science and the advancement of knowledge will somehow be damaged by the growth of human activities in space. As if you can't study something more closely by being there. As if a human being can't do in one day what a robot needs a week to accomplish. As if the exploration and settlement of this new world where I sit today somehow held back the march of scientific progress, rather than driving our advancements and understanding of ourselves and the universe forward at a pace unknown before my ancestors struck out across the sea.

We are truly just at the bare beginnings of the story of humanity and the life forms of the Earth. And we stand poised to take bold steps outwards, if we can do so wisely, economically, and for the right reasons. Those reasons are as wide and varied as those who look at the Moon and stars at night and feel their calling. And this time we can go as a planet, a world reaching beyond itself; its petty borders shattered by the vision of infinity offered in the star filled darkness of space.

And Now the Other Voices

I have laid out some of my ideas in the previous pages, but they are only my own. I could keep going and expanding on the details of each idea, but that is for another time. Now it is time for you to hear other voices. Just as space itself is not the domain of any single reality, the following essays represent no single philosophy-other than that of the Frontier. The Moon and space itself are blank slates, upon which each of us may write our own destiny. And each of us when faced with the challenges offered by this new frontier will draw from what we know, where we come from and how we perceive the universe around us to create our own solutions. And you will see that reflected here.

There is no perfect way to open the frontier, no single answer as to how we should do it. (not even my own!) As I pointed out previously, what we do know is what does not work, as Apollo showed us very clearly. Yet even given that backdrop there are many who would dismiss its lessons and repeat the same failed strategies. In fact, we have seen them fail again and again as this nation and others have poured billions into government human space programs and projects over the last 30-40 years, with little to show for it beyond a few memorable stunts. So the all-government, dead-end, national prestige with a little-bit-of-science approach is one you will not see in this and following volumes. You will see no arguments for multi-billion dollar government programs to go out and "do the Moon and Mars" for us. They simply do not work, and as far as we are concerned this view has been entirely discredited. Even if some of those in our governments have yet to realize it.

What you will see are a wide and varied set of ideas about the why and how of returning to the Moon the right way, all with an eye to our one key premise-This Time We Stay! To us this means all kinds of people, all kinds of governments, all kinds of companies and institutions engaged in the energetic and exciting work of permanently expanding humanity and the life of Earth beyond this planet.

We have chosen a variety of types of essays, and you will hear different voices speaking through them. And I guarantee you that you will not agree with all of them. We begin with some basic and perhaps boring facts about the Moon, but they will be useful to anyone new to this discussion. Then we look into the Big picture, the policies and points of view of some of our space leaders. These include pieces by some of those who were in the White House when the decisions were made to return to the Moon, to others who have been on commissions and groups that have looked into the methods NASA might use to do its part. We look at different aspects of the legal issues surrounding the ownership of property on the Moon. Such issues will loom ever larger, and may turn out to be pivotal, as the first companies decide whether or not to invest in activities such as mining, and the first potential settlers make their choices as to whether it is worth the risk-if in the end they may have nothing to hand down to their children.

We will look into some of the business of space, the concepts and ideas needed to assure that the massive and productive machine of free-enterprise can be successfully unleashed beyond the Earth. Transportation, as I have mentioned, is a key to the opening of the frontier, and to supporting our first lunar communities, so we will look into some of the ideas people have in this area. And what will we be carrying? What will we be doing there? A taste will be offered, although it is only a taste, as there are more possible activities than we could put into this first volume. Towards the end of the book we will look at the Really Big picture, and touch on the spiritual aspects of what it will mean to begin this great quest outwards from the Earth, and why, as human beings, it is up to us to do so. And we shall end with a short piece to tickle your imaginations, by one of our best science fiction writers, as to what the wild, wild Moon frontier might look like in a 150 years or so-if we succeed.

That we must do this great work now and not wait for future generations is the committed position of every one who has had anything to do with this book, from editors to authors, to the artists who did the cover to yes, even the publishers, as Apogee Books has had a history of supporting this great cause. Of course the Space Frontier Foundation is central to this work, and has been the back-bone of this movement for many, many years. We invite you to join us, and if not to join whatever group or company that suites you, and to work in whatever way you can to support the frontier.

So read on, enjoy, agree or disagree with what you read. As do we. And when you are done, go look at the night sky and that silent grey orb, think, and decide what you will write and do to help us to Return to the Moon!

Andrew Chaikin can well be called one of the world's best chroniclers of the American space program. His book, "A Man on the Moon" tells the story of Apollo like no other, and was the basis for the Tom Hanks produced HBO series "From the Earth to the Moon". Andy not only looks to the past, he looks to the future. Although living and breathing as part of the space "establishment", he "gets it" when it comes to a permanent return to the Moon and what will follow. As Andy often points out, we stand on the shoulders of giants. Let us make them proud.

The Next Age of Lunar Exploration

by Andrew Chaikin

I remember when it was just a dream, when going to the Moon was just something depicted in childhood picture books on space travel. A time when astronauts were just characters in movies like *Missile to the Moon,* and TV shows like *Men Into Space.* I remember looking at the Moon through my backyard telescope on hazy Long Island summer evenings, aware that it was a world waiting to be explored. And I remember being aware, too, that I would live to see people travel there.

How fantastic it was to see that exact adventure unfold on the photo pages of *Life* magazine: First, Ed White, his face hidden behind a mirrored visor, floating in the void a hundred miles above the Earth. Next, Gemini 6 had closed to within a foot of Gemini 7, in a miracle of precision. The incandescent fury of the first Saturn V launch, rising from the Earth. And the Earth itself, a blue and white gem suspended in blackness, photographed from lunar orbit during Apollo 8. Finally, on that hot July night in 1969, to sit transfixed before the TV, looking at fuzzy black and white images coming from the Lunar Sea of Tranquillity—a midsummer night's dream come true. To hear the voices of Neil Armstrong and Buzz Aldrin as they walked on that ancient dust, and see them moving like snowmen brought to life. Even then, these things made me feel supremely lucky to be alive.

But the full impact of the adventure was still years away. Not until fifteen years later did I find myself looking skyward and savoring the staggering reality of what I had witnessed. The Moon was no longer just a light in the sky, but a *place*, a place where Americans had actually walked. I hungered to know what that experience was like for the men who made the journey. So began my eight-year effort to become a chronicler of Apollo, to tell the stories of the astronauts who went to the Moon.

That effort gave me a remarkable intimacy with the Moon experience, through my conversations with the astronauts. I remember sitting in Alan Bean's Houston, Texas living room as he described how tricky, and how enjoyable, it was to run in the Moon's one-sixth gravity, how he had to constantly watch out for rocks and craters as he bounded along in slow motion. I sat across from Dave Scott in a restaurant in Marina Del Rey, California as he recalled the view he had high on the slopes of Mount Hadley Delta, as he looked out at a pristine and ancient wilderness; it was a view that literally stopped him in his tracks. During a six-hour conversation in a hotel in Washington, D.C., Buzz Aldrin told me when he walked on the Sea of Tranquillity he noticed the ground curving subtly away from him in all directions; he could actually tell with his own eyes that he and Neil Armstrong were standing on a sphere. And over a seafood lunch in Providence, Rhode Island, Bill Anders confessed that the memory of seeing the Earth as a tiny Christmas tree ornament in the void—that is, being able to validate the Copernican view of the universe just by looking out a window—often had to take second place to the day-to-day concerns of being a corporate executive. Such was the strange duality of Life After the Moon.

I cherish those conversations, and the moments when I could stop being a journalist and savor the feeling of connection with that incredible adventure. The experience of a lunar voyage—even a vicarious one—is a precious gift these men have shared not only with me, but with countless others who heard them talk about Apollo.

Now we have a chance to recapture that gift. The Moon is once again a target for exploration by robots and humans, as part of NASA's Vision for Space Exploration. But there are some who would bypass the Moon in favor of going directly to Mars. They say the Moon will be a distraction; they say the Moon is boring. Their viewpoint mystifies me. Scientifically, our nearest celestial neighbor is priceless; it is the Rosetta Stone for decoding the history of the Solar System. And the fact that six teams of astronauts spent a total of about 12 days living and working on the lunar surface means we've barely begun to explore it.

When it comes to the operational aspects of living and working on other worlds, the Moon is invaluable. Just two and a half days from home, it will be, in the words of science-fiction writers Judy and Gar Reeves-Stevens, a kind of Outward Bound school for planetary explorers. And the Moon offers astronauts a uniquely spectacular setting. After returning from his mission to the Moon's Descartes Highlands, John Young called it "one of the most dazzlingly beautiful places ever visited by human beings." The Moon, it seems, is not only our nearest neighbor, but one of the solar system's crown jewels.

Of course, I understand the lure of the Red Planet; I've felt it since I was five years old. In many ways, Mars is the place that seems, most of all, to bring out the explorer in us. But I remember Dave Scott telling me, "Mars is a *tough job*," and he's right. The difficulties of getting there, of staying healthy during a round-trip voyage lasting perhaps three years, of dealing with any emergencies that arise—all will make the challenges of Apollo seem tame by comparison. Compared to the Moon, getting to Mars will be like the difference between hiking in the foothills and climbing Everest.

And then there is the isolation. Bill Anders said that as far as the Earth seemed from lunar orbit—it was so small that he could have covered it with an outstretched thumb—he knew he had barely left home. From Mars, the Earth will be nothing more than a bright star. And while it took radio transmissions from Earth 1.5 seconds to reach the Moon, it will take many long minutes for them to reach Mars, rendering normal conversation impossible. With all these difficulties in mind, even if things go as well as we could hope, we will likely have to wait another 20 years or more to see humans on the Red Planet.

I don't want it to be that long for us to be able to feel the thrill that I felt during Apollo, of hearing human voices coming to us from another world. And I don't think the "been-there-done-that" crowd appreciates how exciting it will be to step outside at night, look up at the Moon, and say, "there are people living there, and learning how to explore the solar system." Lunar voyages will give us the emotional and spiritual lift we'll sorely need during the long haul to Mars. We have a habit, in our culture, of throwing away valuable things, but this is one experience we can't afford to discard.

And so I can't wait for a new age of lunar exploration. I envision a base at Shackleton crater near the Moon's South Pole, where the view of Earth will hang just above the horizon like something out of Kubrick's *2001*. When it happens, we on Earth will share the adventure more vividly than we ever could during Apollo.

And we'll be able to look up into the deepening dusk, find that bright, cratered world, and know that we are witnessing nothing less than new phase in human evolution. And the Moon will become not just a destination but a catalyst for humanity's transformation into an inter-planetary species.

Andrew Chaikin has authored books and articles about space exploration and astronomy for more than two decades. He is best known as the author of A Man on the Moon: The Triumphant Story of the Apollo Space Program. This acclaimed work was the basis for Tom Hanks' HBO miniseries, From the Earth to the Moon, which won the Emmy for best miniseries in 1998. Chaikin spent eight years writing and researching A Man on the Moon, including hundreds of hours of personal interviews with each of the 23 surviving lunar astronauts. Apollo moonwalker Gene Cernan said of the book, "I've been there. Chaikin took me back." A former editor of Sky & Telescope magazine, Chaikin has also been a contributing editor of Popular Science and has written for Newsweek, Air&Space/Smithsonian, World Book Encyclopedia, Scientific American, and other publications. A graduate of Brown University, Chaikin served on the Viking missions to Mars at NASA's Jet Propulsion Laboratory, and was a researcher at the Smithsonian's Center for Earth and Planetary Studies before becoming a science journalist in 1980.

Courtney Stadd was appointed to shepherd NASA during the transition to the Bush presidency. He has traveled back and forth between NASA, the White House, and private industry many times, and is one of the architects behind our current push into space. When President Bush announced we would be returning to the Moon and begin the human settlement of Mars, I was privileged to be one of only 20 guests in the room invited by the White House. It was a bit odd, given my history as a "radical" in the space field. After the speech, as people milled about, I saw Courtney Stadd standing with a group of government people. As I walked up, he introduced me as "Mr. Moon." At that moment I began to understand that my presence was not an accident, but part of a very strategic effort on his part to blend new ideas and players into the chemistry needed to realize the vision of his former boss, the President. And thus I bring him to this party, as a voice who knows why, as a nation, we must succeed, and what the impediments are. (Ironically, as he points out, the biggest obstacle to establishing ourselves on the Moon and beyond is: *ourselves.)*

Marketplace of Competing Ideas will Determine Alternate Futures

by Courtney A. Stadd

Dear reader, with your forbearance, I would like you to follow me in your imagination to the year 2015. Allow me to suggest a sampling of possible scenarios that might well result from choices made today.

"Hello, I am Bob Johnson, reporting live from the mission control center of the China National Space Agency at the Jiuquan Satellite Launch Center in northwest China's Gansu Province, where we are about to see live video of the first astronaut or, taikonaut, to return to the Moon taking her first step on the lunar surface... of course, this Chinese astronaut represents the first human to step on the lunar surface in 43 years."

"Hello, this is Nancy Adkins reporting live from our Fox "Eye in Space" media platform where Ted Nellis, employee of Amazon.com billionaire Jeff Bezos' Blue Origin's company, is about to deploy the first space hotel, the Trump Celestial Plaza, at Taurus-Littrow, where the last American astronauts walked on the lunar surface in 1972. This project resulted from a deal struck in 2008 between 'The Donald', Mr. Elon Musk, who provided the low cost cargo and crew transport vehicles, and Mr. Robert Bigelow, President of Bigelow Aerospace, who has successfully revolutionized the economics of building and deploying space habits in low Earth orbit and on the lunar surface. In fact, as many viewers may already know, I am reporting from a module, orbiting the earth, built by Mr. Bigelow's company."

"Hi Aaron, this is Miles O'Brien bringing you this special report live from Houston where you can literally hear a pin drop in the room behind me. Every NASA man and woman here is glued to those big monitors you see behind me which are beaming live images of the hatch to the "Glory I" which is about to open and from which the first US astronaut since Gene Cernan and Harrison Schmidt will descend to the lunar surface. And amazingly, Aaron, this mission is taking place roughly on schedule as originally announced by former President George W. Bush, who is here tonight with his family to watch this historic event."

And finally ...

"Welcome to the World In Retrospect where every week we try to take you, the viewer, back in time to understand why certain government initiatives went awry. Tomorrow morning at ten o'clock we will bring you live coverage of hearings by the Joint House and Senate Intelligence Committee. The Committee has called on the nation's Director of Intelligence, Dr. Chelsea Clinton, to respond to new reports of apparent laser weapons tests being conducted on the far side of the Moon by the Chinese-Indian lunar expedition force, and their reported replacement of US flags left by the Apollo astronauts with Chinese and Indian flags to bolster their claim of ownership of the Moon." In light of these hearings, this evening we review the long forgotten decision by former President George W. Bush to commit this nation to go to the Moon and on to Mars and try to understand why the initiative went nowhere and why President Britney Spears and her Administration is not only facing criticism for failing to respond to the actions on the Moon, but is also on the defensive explaining why American aerospace is now considered the least competitive industry sector—far behind Europe and Asia.

Far-fetched though some of these imaginary newscasts may be, this nation—and the world, really—is at a critical turning point in the history of space exploration. The choices made now and the methods used to implement those choices will determine whether the United States continues to have a viable, effective space exploration capability, or whether it will fail to match a renewed vision with a plan of action that discards old ideas of government-first or government-only approaches and instead pursues an aggressive collaboration with the commercial sector.

I view the efforts to return to the Moon and lay the groundwork for permanent human exploration and settlement of the space frontier to be a very serious undertaking and worthy of some of our best and most enterprising technical and business minds. But as any serious student of the history of aviation and space-related technology knows, it is sometimes difficult to determine whether some pioneering entrepreneurial ventures are serious or whether they are ultimately exercises of ego-driven and self-delusional fantasies. Such self-delusion, of course, is not the exclusive domain of the private entrepreneur.

What appears to be delusion may also be a form of scientific arrogance, or the rigid pursuit of a solution based only on known "facts" or familiar principles and assumptions, not imbued with a sense of innovation.

A case in point is Samuel Langley's famous ill-fated, steam powered aerodrome—now on display at the Air & Space Museum's new Hazy Center near Dulles Airport. In 1903, during his tenure as the head of the Smithsonian Institution (at the time this was the equivalent of the nation's Chief Scientist) Dr, Langley was awarded a $50,000 government grant to build what he regarded as the first powered aircraft. The result was a Rube Goldberg contraption, which ultimately suffered an ignominious fate by plunging into the Potomac River. Meanwhile, of course, a few hundred miles to the south, a pair of brothers were making their final preparations to take their self-engineered, self-financed and self-built aircraft on the world's first powered, piloted flight, just a little over two weeks after Langley's aerodrome splashed into the Potomac for the second time.

And how any people scoffed at Alberto Santos-Dumont who traveled around Paris in 1901 in a motor powered one-man dirigible he invented as he went barhopping around the city, circling above the crowds and crashing into rooftops? His biographer, Paul Hoffman, says he was hailed in France as the man who conquered the air. Unfortunately, for Mr. Santos-Dumont, word of the first flights of Orville and Wilbur had not yet reached Europe.

Some narrow-minded people might write such stunts off as semi-relevant footnotes to the history of the real achievements in aviation and space travel. I, for one, prefer to view them as wonderful examples of what happens when humanity's inventive spirit is allowed to blossom in the marketplace of competing ideas and innovation. (Of course, the Wright Brothers may have rightly wondered if Mr. Langley's grant of $50,000 from Uncle Sam was provided on a fair and open basis!)

I am hopeful that I will live long enough to see the emergence of a robust marketplace of magnificent men (and women) and their magnificent machines and visions for travel in the space field—as robust as the one that existed in the early days of aviation. Unfortunately, I fear we are still quite a ways from such a vision becoming reality.

I think the President's Vision, however, was a good first step to helping foster such a marketplace. It lays out a vision for the civil space community that identifies measured, pragmatic, evolutionary steps for space exploration. Achievement of those goals will require a number of ambitious capabilities to be developed and demonstrated on a timeline that is equally ambitious. It will also require substantial changes and redirection of NASA leadership and practices, vis-à-vis the private sector and the proper dividing line between exploration and operations of space-based assets.

Based on my personal experience, I know that it is nearly impossible to craft a national policy that satisfies all the various and sundry stakeholders. There will always be "rice bowls", or those who resist change when new priorities are set, and have a vested interest in preserving certain NASA programs that will be terminated or redirected as a result of the new vision. There will also be those that are frustrated by what they may view as an overly deliberate, evolutionary approach to realizing the President's goals.

Some will feel that such an approach is inadequate to address what they view as a new kind of race for space. The original space race was bipolar, with players and issues that were starkly differentiated for the whole world to see. This time, the players are many, the issues are complex and nuanced. But what is at stake is nothing less than dominance in a field that has profound implications for leadership in the global technology marketplace. If a nation fails to maintain leadership in semi-conductors it is not good but it is not devastating. But handing over leadership in space has profound national security (literally losing the high frontier) and commercial implications.

In the aftermath of the Columbia accident, it was widely stated that NASA had lost its vision, and needed a new mandate to find its way back to prominence and credibility as a forward-looking and forward-moving agency. President Bush provided the essence of that vision, but also set the stage for an even greater debate on what NASA's role should be in the broad scheme of exploring and exploiting space. The vision articulated by the President is something that must be sustained beyond the time that he will be President. It is thus larger than the Administration, larger than NASA, and must, indeed, be a national vision. Already, it is obvious that NASA cannot be the same agency it was before the vision was announced. It cannot do all the things it was doing before and still do all that is required to advance the new exploration agenda.

Traditional partners with NASA and programs within NASA across a broad spectrum are seeing funding decreased, programs cancelled or reformulated to more directly support the exploration requirements. Members of Congress are expressing concern about the loss of constituent jobs and the elimination of important national research capabilities. These pressures and others will come increasingly in conflict with the new exploration efforts in NASA, and unless alternative means are found outside of NASA's limited resources to either assume some of those traditional activities or to conduct elements of the new exploration activities, or both, it is almost certain to mean an early death for the new vision for exploration. The kind of adjustments needed are not limited to "supporting services" provided by private sector interests, but extend to such things as transitioning major current activities to private sector management, and perhaps even ownership. There will be increasing pressure to build out the space station to the promised level of research capacity that served as the basis for the political support for the space station over the past sixteen years. Yet, it is already clear that it is not, and will not, be possible within the resources available, for NASA to meet that objective. The answer is to either redefine the "completion" level for the space station, which must be done in concert with the leadership of partner nations, or to devise a means of relieving NASA of the enormous burden of sustaining space station operations through some sort of arrangement whereby the private sector can assume responsibility for the space station operations in a manner which allows them to benefit—and profit—from that undertaking. To do that, of course, will require new kinds of contractual approaches, which will require new kinds of legal authorities from the Congress and force the adoption of an entirely different attitude within NASA regarding the private sector role in exploration. Something that is, in my view, way overdue!

Even without such a transition of space station operations and utilization, the need exists for alternative human-rated vehicles that can deliver crews and supplies to the space station and return payloads to Earth. With direct private sector involvement in the space station operations, the burden-sharing options for development of new vehicle capabilities are likely to increase, which could further relieve budgetary pressure on NASA. Such a sea-change in the role and mission of NASA and its relationship with the private sector is something that, in my view, must be considered if the vision for exploration is to succeed and survive the natural pressures already expressing themselves to sustain many non-exploration activities that NASA has historically undertaken, especially in the fields like aeronautics, fundamental life sciences research and Earth science.

My initial sense is that some of the senior and mid-level NASA officials, involved with the new exploration initiative, are making best efforts to work with non-traditional players in leveraging and expanding the investment of taxpayer dollars in the space program. But many times they encounter resistance, for example, from the legal and procurement program offices who are largely staffed with people who resist change and innovation. Of course, NASA is in the midst of an ongoing effort to persuade the political system that it is putting in place the checks and balances to guard against its historical propensity to significantly underestimate the costs and technical challenges associated with achieving ambitious goals. In the absence of achieving such credibility, the agency will continue to struggle to obtain the resources needed to carry out its new mandate. And, frankly, I think this is a major reason—aside from some tactical budgetary victories—for the current inability of the new initiative to gain significant traction within Congress.

Some members of the NASA leadership team are also to be applauded for their openness to new initiatives such as the NASA Centennial Challenges Program—inspired by the X-Prize model. It proposes to identify $20 million in a series of annual prizes for revolutionary, breakthrough accomplishments from innovators not usually affiliated with the space program. Examples of potential candidate programs include nano-materials, very low cost robotic space missions and spacecraft power systems. The key to this program's success, however, is to ensure minimal bureaucratic intrusion and efforts by "rice bowls" to vector the resources into programs that perpetuate the status quo versus truly advancing unorthodox inventions and ideas. Done right, it could make a major contribution to fostering a paradigm shift in how the Agency works with the private sector.

In the area of relationship building with the private sector, allow me to make the understatement of the year: NASA suffers from a major lack of credibility with the private sector. Over the years, the agency has generated various initiatives that purported to promote greater involvement by non-traditional players with the space program only to see the entrepreneurs disappointed by the failure of the agency to effectively follow-through. As a result, the private sector will be following this Centennial Challenges' initiative with a mix of great hope and significant skepticism.

As we witness the emergence of a new generation of space entrepreneurs seeking to ride the wave of this new vision, it is useful to note that the nation's pension funds, banks, and insurance companies appear to be beginning to re-energize their private equity and debt investments into venture and other forms of capital management. And venture firms are showing signs of a penchant for many of the nano-technology, life sciences, power sources, power technologies and other fundamental technical areas required for support of new space exploration missions. With that said, it is still tough acquiring outside capital for ventures that have "space" as a prefix in their corporate name.

This is all the more reason to be grateful for the entry into the space field of successful high net worth individuals, such as Robert Bigelow, President, Bigelow Aerospace, who is deploying his own significant private resources to build low cost expandable space habitats in low earth orbit with potential applications for use on the lunar surface. Pioneers, such as Bigelow, are demonstrating that the private sector can potentially augment the government's efforts to open the space frontier for the full expression of the human enterprise. It is my firm belief that the success and realization of his efforts, and those of others like him, will serve as a bell-weather that will spark the involvement of a wide array of new investors and partners in what will be seen as a real and viable new "market in the stars."

Speaking of augmenting and leveraging national space initiatives, recent history has seen unprecedented efforts by multiple countries to send probes to the Moon. Examples include:

The European Space Agency's Small Missions for Advanced Research in Technology or SMART-I has already completed its primary technology demonstration mission and has been extended an additional year to enable high resolution mapping of both north and south hemispheres of the Moon, providing surveys of lunar resources, terrain maps for rovers and the like. ESA is planning a series of landers as their encore.

Japan is planning on launching Lunar A and B missions sometime this decade. These missions will study the composition of the Moon.

India is planning a lunar probe for the latter part of the decade that will provide three-dimensional high resolution mapping of the surface in X-ray, near infrared and low energy gamma ray imagery.

And it is not exactly a secret that China is planning a multi-phased strategy for lunar exploration—robotic and perhaps ultimately involving humans

One has to believe that these missions will lead to knowledge of the Moon that would be invaluable to our present day Lewis and Clarks looking to chart their own entrepreneurial way to the Moon. But without a commensurate US effort, that knowledge may be denied to US-based companies and individuals, and will, instead, advance the interests of those who wish to replace US leadership in space exploration with their own and reap the vast rewards that are certain to flow from exploration beyond low earth orbit.

Let me conclude with some random observations on some critical issues we all need to be grappling with as a community:

For those who believe we need to build a heavy lifter (of course we used to have one of those with a 100% success rate, but we ever so cleverly managed to throw away the drawings) some believe we are talking on the order of $10-15B. Certainly that would be the cost if we rely on NASA accounting. A shuttle derivative would hardly be less and might be actually more. So, with money tight, how can we do exploration with what we have in the inventory and without breaking the proverbial bank?

NASA is currently stuck in LEO. How can one most effectively bring the commercial sector along in lock step with the exploration mission? Certainly one key element is building the infrastructure that can be operated cheaply (hopefully by private concerns).

The maximum reuse of space technology is key. For example, we have to avoid building highly specialized space vehicles. Common environmental control and life support systems, structures, power systems, etc. will be needed to form building blocks. What concrete steps are needed to ensure greater reliance on innovation and commercial sources?

Human exploration is built upon a base of robotic exploration. So we need to exploit terrestrial robotic systems and adapt them to the exploration vision. How do we ensure that NASA is a user rather than a developer of advanced technology, in this case robotics? In my view, if NASA has to create everything from whole cloth the whole vision is doomed to failure.

Few of us alone have definitive answers to these and many related questions. Only by working together, and in conjunction with NASA, as appropriate, can the fledgling entrepreneurial space community hope to formulate the necessary real and pragmatic solutions to these and related concerns. Breaking out of the earth's gravity and heading toward another planetary surface is non-trivial, but I believe that we have a real opportunity to foster an environment where the future Wright Brothers and Alberto Santos-Dumonts of space can be given a chance to demonstrate their spirit of inventiveness in exploring the high frontier.

America's space entrepreneurs, who reside in both small and large companies, are poised once again to bring the promise of space to fruition. Frankly, a major challenge is whether the U.S. Government will follow-through on the promise of the new policy vision and foster a truly entrepreneurial environment that will ultimately determine which of the scenarios described at the beginning of this commentary will come to pass.

I am hopeful that NASA may yet find its way to effectively contribute to the nations exploration efforts without co-opting or hamstringing the emerging private sector, entrepreneurial efforts of today's new space pioneers. I believe the new Administrator and his management team can and will enter into the type of innovative partnerships with the commercial sector that maximizes the chances that we might one day witness CNN space reporter Miles O'Brien reporting the landing of the first American on the Moon since Apollo 17 roughly in accord with the milestones laid out by President Bush on January 14, 2004. I would go a step further and propose the real possibility that, if done right, that historic return mission might well be undertaken by the American commercial sector as a co-equal partner with NASA.

Courtney Stadd headed the NASA Transition Team for President Bush. He was then appointed Chief of Staff and White House Liaison, NASA HQ. In July 2003, he left the government to return to the private sector as a management consultant for aerospace-and high tech-related firms.

Each of the essays in this book takes a slightly (or in some cases radically) different approach to the how and why of returning to the Moon. Yet, as you will note, certain themes re-occur frequently. I have left some of these in place, as each author states them a bit differently, and from different political and professional points of view. And frankly, I feel they bear repetition. For his part, Dr. Spudis was chosen to sit on the Aldridge Commission that was set up by President Bush to lay out the basic concepts to be incorporated in the realization of his Vision for Space Exploration, which included permanent human presence beyond Earth orbit, to include the Moon and Mars. A champion of the Moon, Dr. Spudis and his allies have made a strong case within the NASA aerospace establishment for the Moon to be not just a "touch and go" pit stop on the way to Mars—which is how many in the agency had interpreted his words. In this essay Paul offers hi overview of the challenge and rewards that a strong Earth-Moon infrastructure offers in the context of the President's vision.

The Moon: A New Destination in Space for America

by Dr. Paul D. Spudis

On January 14, 2004, President George W. Bush visited the Headquarters of the National Aeronautics and Space Administration in Washington DC to announce a new vision for America's space program. Prominent among the many goals and objectives outlined that day was a return to the Moon to both explore and use it. Although we conducted our initial visits there over 30 years ago, recent important discoveries indicate that a return to the Moon offers many advantages and benefits to the nation. In addition to being a scientifically rich object for study, the Moon offers abundant material and energy resources, the feedstock of an industrial space infrastructure. Once established, such an infrastructure will revolutionize space travel, assuring us continuous, routine access to cislunar space (i.e., the space between and around Earth and Moon) and beyond. The value of the Moon as a space destination has not escaped the notice of other countries—at least four new robotic missions are being flown or prepared for flight by Europe, India, Japan, and China and advanced planning for human missions in many of these countries is already underway.

The Moon is close, accessible with existing systems, and has resources that we can use to create a true, economical space-faring infrastructure

The Moon is a scientific and economic treasure trove, reachable with existing systems and infrastructure that can revolutionize our national strategic and economic posture in space. The dark areas near the poles contain significant amounts (at least 10 billion tons) of hydrogen, most probably in the form of water ice. This ice can be mined to support human life on the Moon and in space and to make rocket propellant (liquid hydrogen and oxygen). Moreover, we can return to the Moon using the existing infrastructure of Shuttle-derived launch systems for only a modest increase in the space budget.

The "mission" of this program is to go to the Moon to learn how to use off-planet resources to create new capability and to make space flight easier and cheaper in the future. Rocket propellant made on the Moon will permit routine access to cislunar space by both people and machines, vital to the servicing and protection of national strategic assets and for the repair and refurbishing of commercial satellites. The availability of cheap propellant in low Earth orbit would completely change the way engineers design spacecraft and the way companies and the government think about investing in space assets. It would serve to dramatically reduce the cost of space infrastructure to both government and the private sector, thus spurring economic investment (and profit).

The Moon is a unique scientific resource on which important research, ranging from planetary science to astronomy and high-energy physics, can be conducted.

Generally considered a simple, primitive body, the Moon is actually a small planet of surprising complexity. Moreover, the period of its most active geological evolution, between 4 and 3 billion years ago, corresponds to a "missing chapter" of Earth history. The processes that work on the Moon—impact, volcanism, and tectonism (deformation of the crust)—are the same ones that affect all of the rocky bodies of the inner solar system, including the Earth. Because the Moon has no atmosphere or running water, its ancient surface is preserved in nearly pristine form and its geological story can be read with clarity and understanding. Because the Moon is Earth's companion in space, it retains a record of the history of this corner of the Solar System, vital knowledge unavailable on any other planetary object.

Of all the scientific benefits of Apollo, an appreciation of the importance of impact, or the collision of solid bodies, in planetary evolution must rank highest. Before we went to the Moon, we had to understand the physical and chemical effects of these collisions, events completely beyond the scale of human experience. Of limited application at first, this new knowledge turned out to have profound consequences. We now believe that large-body collisions periodically wipe out species and families on Earth, most notably, the extinction of dinosaurs 65 million years ago. The telltale residue of such large body impacts in Earth's past is recognized because of knowledge we acquired about the impact process from the Moon. Additional knowledge still resides there; while the Earth's surface record has been largely erased by the dynamic processes of erosion and crustal recycling, the ancient lunar surface retains this impact history. When we return to the Moon, we will examine this record in detail and learn about its evolution as well as our own.

Because the Moon has no atmosphere and is a quiet, stable body, it is the premier place in space to observe the universe. Telescopes erected on the lunar surface possess many advantages. The Moon's level of seismic activity is orders of magnitude lower than that of Earth. The lack of an atmosphere permits clear viewing; with no spectrally opaque windows to contend with, the entire electromagnetic spectrum is visible from the lunar surface. Its slow rotation (one lunar day is 708 hours long, about 28 terrestrial days) means that there are long times of darkness for observation. Even during the day, brighter sky objects are visible through the reflected surface glare. The far side of the Moon is permanently shielded from the din of electromagnetic noise produced by our industrial civilization. There are areas of perpetual darkness and sunlight near the poles. The dark regions are very cold, only a few tens of degrees above absolute zero and these natural "cold traps" can be used to passively cool infrared detectors. Thus, telescopes installed near both poles can see entire celestial hemispheres at once with infrared detectors, cooled for "free," courtesy of the cold traps.

Recent suggestions that lunar dust poses unsolvable problems and difficulties for telescopes on the Moon are incorrect; dust does not "coat" surfaces if left undisturbed. The Apollo astronauts became covered in dust because in some cases, they fell, knelt, or had to literally wallow in dust to pick up the samples they wanted to return. The best evidence that lunar dust creates no long-term problems comes from the performance of the Laser Ranging Retroreflectors (LRRR), which were deployed by Apollo astronauts at four different sites. These passive arrays of glass cubes are used as mirrors to reflect laser pulses sent from Earth in order to precisely measure the Earth-Moon distance. After over 30 years of continuous use and exposure to the lunar dust environment, they show no degradation of photon return whatsoever.

The Moon possesses the resources needed to create a spacefaring transportation infrastructure in cislunar (Earth-Moon) space.

The return of the Apollo samples taught us the fundamental chemical make-up of the Moon. It is a very dry, chemically reduced object, rich in refractory elements but poor in volatile elements. The composition of the Moon is rather ordinary, made up of common Earth minerals such as plagioclase (an aluminum, calcium silicate), pyroxene (a magnesium, iron silicate), and ilmenite (an iron-titanium oxide). The Moon is approximately 40% oxygen by weight. Light elements, including hydrogen and carbon, are present, but in small amounts—in a typical mare soil, hydrogen makes up between 50 and 90 parts per million by weight. Soils richer in titanium appear to be also richer in hydrogen, thus allowing us to infer the extent of hydrogen abundance from the global titanium concentration maps returned by both the Clementine and Lunar Prospector missions.

As usable commodities, lunar materials offer many possibilities. Because radiation is a serious problem for human spaceflight beyond low-Earth orbit, the simple expedient of covering surface habitats with soil can protect future inhabitants from both galactic cosmic rays and even solar flares. Lunar soil can be sintered by microwave into very strong building materials, including bricks and anhydrous glasses that have strengths many times that of steel. When we return to the Moon, we will have no shortage of useful building materials.

Because of its high abundance in the Moon, oxygen is likely to be an important early product. The production of oxygen from lunar materials is not magical, but simply involves breaking the very tight chemical bonds between oxygen and various metals in minerals. Many different techniques to accomplish this task have been developed; all are based on common industrial processes easily adapted to use on the Moon. Besides human life support, the most important use of oxygen in its liquefied form is to make rocket fuel oxidizer. Coupled with the extraction of solar wind hydrogen from the soil, this processing can make rocket fuel the most important commodity of a new lunar economy.

The Moon has no atmosphere or global magnetic field so the solar wind (the tenuous stream of gases emitted by the Sun, mostly hydrogen) is directly implanted onto lunar dust grains. Although solar wind hydrogen is present over most of the Moon in very small quantities, it too can be extracted from soil. Soil heated to about 700° C releases more than 90% of its adsorbed gases. Such heat can be obtained from collecting and concentrating solar energy using focusing mirrors, a readily available form of energy. Collected by robotic processing rovers, solar wind hydrogen can be harvested from virtually any location on the Moon. Additionally, recent discoveries suggest that special areas exist where this material is present in much greater abundance, making its collection and use much easier.

Hydrogen, probably in the form of water ice, at the poles can be extracted and processed into rocket propellant and life-support consumables

The joint DoD-NASA Clementine mission was flown in 1994. Designed to test sensors developed for the Strategic Defense Initiative (SDI), Clementine was an amazing success story. This small spacecraft was designed, built, and flown within the short time span of 24 months for a total cost of about $150 M (FY 2003 dollars), including the launch vehicle. Clementine made global maps of the mineral and elemental content of the Moon, mapped the shape and topography of its surface with laser altimetry, and gave us our first good look at the intriguing and unique polar regions. Clementine did not carry instruments specifically designed to look for water, but an ingenious improvisation used the spacecraft communications antenna to beam radio waves into the polar regions; radio echoes were observed using the Deep Space Network dishes. Results indicated that material with reflection characteristics similar to ice is found in the permanently dark areas near the south pole. This major discovery was subsequently supported by a different experiment flown on NASA's Lunar Prospector spacecraft in 1998 that indicated large amounts of hydrogen are to be found near both poles.

The Moon contains no internal water; water is added to it over geological time by the impact of comets and water-bearing asteroids. Because the Moon's axis of rotation is nearly perpendicular to the plane of the ecliptic (the plane in which Earth and Moon orbit the sun), the sun always appears on or near the horizon at the poles. If you're in a hole, you never see the sun; if you're on a peak, you always see it. Depressions near the poles never receive sunlight; these dark areas are very cold, only a few degrees above absolute zero. Any water that gets into these polar "cold traps" cannot get out and over time, significant quantities can accumulate. Our current best estimate is that over 10 billion cubic meters of water exist at the poles, an amount equal to the volume of Utah's Great Salt Lake—without the salt! Although hydrogen and oxygen can be extracted directly from the soil as described above, such processing is difficult and energy-expensive. Polar water has the advantage of already being in a concentrated useful form, greatly simplifying scenarios for lunar return and habitation. Broken down into hydrogen and oxygen, water is a vital substance both for human life support and rocket propellant. Water from the lunar cold traps advances our space-faring infrastructure by creating our first space "filling station."

The poles of the Moon are useful from yet another resource perspective—the areas of permanent darkness are proximate to areas of near-permanent sunlight. We have identified several areas near both the north and south poles that offer near-constant illumination by the sun. Moreover, such areas are in darkness for short periods, interrupting longer periods of illumination. Thus, an outpost or establishment in these areas will have the advantage of being in sunlight for the generation of electrical power (via solar cells) and in a benign thermal environment (because the sun is always at grazing incidence); such a location never experiences the temperature extremes found on the equator (from 100° to –150° C). The poles of the Moon are inviting "oases" in near-Earth space.

Current launch systems, infrastructure, and hardware can be adapted to this mission and we can be back on the Moon for only a modest increase in existing space budgets.

America built the mighty Saturn V forty years ago to launch men and machines to the Moon in one fell swoop. Indeed, this technical approach was so successful, it has dominated the thinking on lunar return for decades. One feature of nearly all architectures of the past twenty years is the initial requirement to build or re-build the heavy lift launch capability of the Saturn V or its equivalent. Parts of the Saturn V were literally hand-made, making it a very expensive spacecraft. Development of any new launch vehicle is an enormously expensive proposition. What is needed is an architecture that accomplishes a lunar return with the least amount of new vehicle development possible. Such a plan will allow us to concentrate our efforts and energies on the most important aspects of the mission—learning how to use the Moon's resources to support space flight beyond low Earth orbit.

The plan for lunar return devised by the Office of Exploration at the Johnson Space Center has several advantages. It uses Shuttle-derived components, augmented by existing expendable boosters, to deliver the pieces of the lunar spacecraft to Earth orbit—lander, habitat, and transfer stage. Assembled into a package in space, these items are then transferred to a point about 4/5 of the way to the Moon, the Moon-Earth Lagrangian point 1 (L1). The L1 point orbits the Earth with the Moon such that it appears "motionless" to both bodies. Because there is no requirement for quick transit, cargo elements can take advantage of innovative technologies such as solar electric propulsion and weak stability boundaries between Earth, Sun, and Moon to make long, spiraling trips out to L1, thus requiring less propellant mass. These unmanned cargo spacecraft can take several months to get to their destinations. The habitat module can be landed on the Moon by remote control, activated, and await the arrival of its occupants from Earth.

The crew is launched separately in the new Crew Exploration Vehicle and uses a chemical stage and a quick transfer trajectory to reach the L1 depot in a few days. They transfer to the lander, descend to the surface, and conduct the surface mission. As mentioned above, the preferred landing site is an area near one of the Moon's poles; the south pole is most attractive from the perspective of both science and operations. The goal of our mission is to learn how to mine the resources of the Moon as we build up surface infrastructure to permit an ever-larger scale of operations. Thus, each mission brings new components to the surface and the size and capability of the outpost grows over time. Most importantly, the use of lunar-derived propellants means that more than 80% of the spacecraft weight on return to Earth orbit need not be brought from Earth. A properly designed mission will return to Earth not only with sufficient fuel to take the craft back to the Moon for another run, but also to provide a surplus for sale in low Earth orbit. It is this act that creates the Earth-Moon economy and demonstrates a positive return on investment.

In addition to its technical advantages, this architecture offers important programmatic benefits. It does not require the development of a new heavy lift launcher, always a high cost item in any new space mission. The use of the L1 point as a staging depot allows us to wait for proper alignments of the Earth and Moon; the energy requirements to go nearly anywhere beyond this point are very low. The use of newly developed, low-thrust propulsion (i.e., solar-electric) for cargo elements drives new technology development. We will acquire new technical innovation as a by-product of the program, not as a critical requirement of the architecture.

The importance of using Shuttle-derived launch vehicles and commercial launch assets in this architecture should not be underestimated. Costs in space launch are almost completely dominated by the costs of people and infrastructure. To create a new launch system requires new infrastructure, new people, new training. Such costs can make up significant fractions of the total program. By using existing systems, we can concentrate our resources on new equipment and technology, focused on the goal of finding, characterizing, processing, and using lunar resources as soon as possible.

A return to the Moon gives the nation a challenging mission and creates capability for the future, by allowing us to routinely travel at will, with people, throughout the Earth-Moon system.

Mining the Moon for propellant establishes a robust transportation infrastructure, capable of delivering people and machines throughout cislunar space. Make no mistake—learning to use the resources of the Moon or any other planetary object will be a challenging technical task. We must learn to use machines in remote, hostile environments, working under difficult conditions with ore bodies of small concentration. The unique polar environment, with its zones of near-permanent illumination and permanent darkness, provides its own challenges. But for humanity to have a future in space, we must learn to use the materials available off-planet. We are fortunate that the Moon offers us a nearby, "safe" laboratory to take our first steps in using space resources. Initial blunders in operational tactics or feedstock processing are better practiced at a location three days from Earth than from one many months away.

A mission learning to use these resources is scalable in both level of effort and the types of commodities to be produced. We begin by using the resources that are the easiest to extract. Thus, a logical first product is water derived from the polar deposits. Water can be produced here regardless of the nature of the polar volatiles—ice of cometary origin is easily collected and purified, but if the polar materials are composed instead of molecular hydrogen, this substance can be combined with oxygen extracted from rocks and soil (through a variety of processes) to make water. Water is easily stored and used as a life-sustaining substance for people or retained in its separate state for its use as rocket propellant.

Although we possess some information to plan a lunar return now, investment in a few robotic precursor missions would reduce programmatic risk and enhance initial return. We should map the polar deposits of the Moon from orbit using imaging radar to "see" the ice in the dark regions. Such mapping would establish the details of ice locations and its thickness, purity, and physical state. The next step should be to land small robotic probes to conduct chemical analyses of the polar deposits. Although we expect water ice to dominate the deposit, cometary cores are made up of many different substances, including methane, ammonia, and organic molecules, all of which are potentially useful resources. We need to inventory these species and determine their chemical and isotopic properties and their physical nature and local environment. Just as the way for Apollo was paved by such missions as Ranger and Surveyor, a set of robotic precursor missions, conducted in parallel with the planning of the manned expeditions, can make subsequent human missions safer and more productive.

After the robotic missions have documented the nature of the deposits, focused research should be undertaken to develop the machinery needed to be transported to the Moon as part of the human expedition. Then human-tended mining processes and extraction principles will be established and validated, thus paving the way to commercialization of the production of lunar hydrogen and oxygen.

This mission creates routine access to cislunar space for people and machines, which directly relates to important national economic and strategic goals.

By learning space survival skills close to home at the Moon, we create new opportunities for exploration, utilization, and wealth creation. Space is no longer a hostile place that we tentatively visit for short periods; it becomes instead a permanent part of our world. Achieving routine freedom of cislunar space makes America more secure (by enabling larger, cheaper, and routinely maintainable assets on orbit) and more prosperous (by opening an essentially limitless new frontier.)

As a nation, we rely on a variety of government assets in cislunar space—weather satellites, GPS systems, and a wide variety of reconnaissance satellites. In addition, commercial spacecraft continue to make up a multi-billion dollar market, providing telephone, Internet, radio and video services. America has invested billions in this space infrastructure. Yet at the moment, we have no way to service, repair, refurbish or protect any of these spacecraft. They are vulnerable to severe damage or permanent loss. If we lose a satellite, it must be replaced. From redesign though fabrication and launch, such replacement takes years and involves extraordinary investment in the design and fabrication so as to make them as reliable as possible.

We cannot access these spacecraft because it is not feasible to maintain a human-tended servicing capability in Earth orbit—the costs of launching orbital transfer vehicles and propellant would be excessive (it costs over $10,000 to launch one pound to low Earth orbit using the Shuttle). Creating the ability to refuel in orbit, using propellant derived from the Moon, would revolutionize the way we view and use our national space infrastructure. Satellites could be repaired, rather than abandoned. Assets can be protected, rather than written off. Very large satellite complexes could be built and serviced over long periods, creating new capabilities and expanding bandwidth (the new commodity of the information society) for a wide variety of purposes. And along the way, we create new opportunities and make discoveries.

A return to the Moon, with the purpose of learning to mine and use its resources, thus creates a new paradigm for space operations. Space becomes part of America's industrial world, not an exotic environment for arcane studies. Such a mission ties our space program to its original roots in making us more secure and more prosperous. But it also enables a new and broader series of scientific and exploratory opportunities. If we can create an infrastructure to routinely access cislunar space, we have a system that can take us to the planets.

Timing is everything: It is important for America to undertake this mission NOW, rather than later.

Many nations have recently indicated an interest in the Moon. The possible collection and use of lunar resources raises some interesting political and economic issues. Currently, the 1967 United Nations Treaty on the Peaceful Uses of Outer Space prohibits claims of national sovereignty on the Moon or any other object. However, it is not clear that private claims are likewise prohibited under this treaty. The 1984 United Nations Moon treaty specifically prohibits private ownership of lunar assets, but the United States, Russia, and China are not signatories to that treaty, ratification of which was specifically rejected by the United States Senate.

Our initial return to the Moon would be an engineering and scientific research and development project. We undertake our studies of lunar resource processing to ascertain whether such resources can indeed be extracted and used. Our presence on the Moon does not give us title to it. However, a strong and continuing American presence there can help establish *de facto* the broad legal framework and economic paradigm of democratic, free-market capitalism off the Earth. It is not clear that other nations would be similarly inclined. In short, regardless of impressions to the contrary, we are indeed in a race to the Moon—a race between competing economic and political paradigms and one no less important than the earlier "space race" in establishing and maintaining global stability. History has shown that only our system produces the most wealth and freedom and highest quality of life for the most people in the shortest time. America needs to continue to lead in space, ensuring that an open economy and a self-determining, democratic framework is established off-Earth.

The infrastructure created by a return to the Moon allows us to travel to the planets in the future more safely and cost effectively.

This benefit comes in two forms: first, developing and using lunar resources enables flight throughout the Solar System by permitting fueling of the interplanetary craft with materiel already in orbit, saving the enormous costs of launch from Earth's surface. Second, the processes and procedures that we learn on the Moon are lessons that will be applied to all future space operations. To successfully mine the Moon, we must learn how to use machines and people in tandem, each taking advantage of the other's strengths. The issue isn't "people or robots?" in space; it's "how can we best use people and robots *together* in space?" People bring the unique abilities of cognition and experience to exploration and discovery; robots possess extraordinary stamina, strength, and sensory abilities. We can learn on the Moon how to best combine these two complementary skill mixes to maximize our exploratory and exploitation abilities.

Returning to the Moon will allow us to regain operational experience on another world. Our activities there make future planetary missions less risky because we gain this valuable experience in an environment close to Earth, yet on a distinct and unique alien world. Systems and procedures can be tested, vetted, revised and re-checked. Exploring a planet is a difficult task to tackle green; learning to live and work on the Moon gives us a chance to crawl before we have to walk in planetary exploration and surface operations.

The Moon provides a nearby laboratory and industrial test-bed where we can hone our exploratory skills and lay the foundations for a future space-based economy. Human expansion to the Moon will provide new opportunities and horizons for the American entrepreneur, business, and workforce. Developing new technologies has always led to new markets and increased our prosperity. Expansion of the economy is vital to our national health and security.

America needs a challenging, vigorous space program. It must present a mission that inspires, educates, and enriches. It must relate to important national needs yet push the boundaries of the possible. It must serve larger national concerns beyond pure science. The President's program fulfills these goals. It is a technical challenge to the nation. It creates security for America by assuring access and control of our assets in cislunar space. It creates wealth and new markets by producing commodities of great commercial value. It stimulates and inspires the next generation by example.

A return to the Moon by America is a giant step into the Solar System.

Paul D. Spudis is a Principal Professional Staff member at the Johns Hopkins University Applied Physics Laboratory in Laurel, Maryland and Visiting Scientist at the Lunar and Planetary Institute in Houston, Texas. He is a geologist who received his education at Arizona State University (B.S., 1976; Ph. D., 1982) and at Brown University (Sc.M., 1977). Since 1982, he has been a Principal Investigator in the Planetary Geology and Geophysics Program of the NASA Office of Space Science, Solar System Exploration Division, specializing in research on the processes of impact and volcanism on the planets. Spudis has been a member of the Committee for Planetary and Lunar Exploration (COMPLEX), an advisory committee of the National Academy of Sciences, and the Synthesis Group, a White House panel that in 1990-1991, analyzed a return to the Moon to establish a base and the first human mission to Mars. He was Deputy Leader of the Science Team for the Department of Defense *Clementine* mission to the Moon in 1994 and is the Principal Investigator of an imaging radar experiment on the Indian Chandrayaan-1 mission, to be launched to the Moon in 2007. He was a member of the President's Commission on the Implementation of U. S. Space Exploration Policy, whose report was issued June, 2004 and in September 2004, was presented with the NASA Distinguished Public Service Medal for his work on that body. He is the recipient of the 2006 Von Karman Lectureship in Astronautics, awarded by the American Institute for Aeronautics and Astronautics. He is the author or co-author of over 150 scientific papers and three books, including *The Once and Future Moon*, a book for the general public in the Smithsonian Library of the Solar System series, and (with Ben Bussey) *The Clementine Atlas of the Moon*, published in 2004 by Cambridge University Press.

Dr. Yoji Kondo is one of those people who fits the mold of a space frontier renaissance man. A brilliant astronomer, he is also a writer of science fiction and one of the nation's top Aikido Masters. In fact Yoji so impressed the late Robert Heinlein that he shows up in at least two of his novels. For many years he has been a quiet voice speaking to me of the need for our movement, the need to try whenever possible to work with the system to change it, and the understanding that at times the system needs to be circumvented. In this essay Dr.Kondo dives into one of the great debates of the old pre-frontier space program; Why send people when robots will do? He does so with one foot in the old paradigm and one foot in the new, yet he makes the case well. This debate was one of the great issues when going into space was only about science, but to us that is only one reason we want to go. Space is a frontier, it is a frontier to be opened and settled by human beings and the life we bring with us. Toasters don't breed.

Manned and Robotic Explorations of Space:

Two aspects of the same basic human drive to explore the unknown.

by Dr. Yoji Kondo

Since the days of the Apollo missions to place American astronauts on the Moon, there have been active debates among various professionals, politicians, and others concerning the relative merits of human space exploration and scientific research in space using robotic or automated instruments. These debates have sometimes gotten so heated that they have become acrimonious from time to time.

During the early years of the space age, those working in the human space program maintained that human presence in space was critical to exploring the final frontier. Much of the public appeared to share that same sense of enthusiasm about humankind's future in space. Neil Armstrong's remark as he became the first man to walk on the Moon, "That's one small step for man, one giant leap for mankind," was an inspiration to millions of people; it proclaimed that the golden age of space exploration had arrived—and that people, not robots, were going to lead humanity on the adventure.

At the same time, there were people who maintained that we should spend most, if not all, of our resources on "pure" scientific research projects in space using robotic and automated instruments, such as orbiting astronomical observatories and geophysical satellites. These people believed that robotic envoys were virtuous on their own merits, as they pursued only pure scientific goals, and would provide a far greater scientific return for the investment than human explorers would. Still others did not much like the idea of spending funds on a space program of any sort. Some astronomers, for example, insisted that ten five-meter telescopes would cost as much as launching one orbiting astronomical observatory. Despite the allegations made by some people about 'all that money spent in space', the funds on the exploration of space have all been expended right here on this planet. [Not a single job has been outsourced to ETs, either.]

Are human exploration of space and robotic probes in space necessarily diametrically opposite?

The federal government will always have to make decisions about how to allocate the available public funds for the space program between human and robotic programs; such discussions are a necessary part of the process for rational distributions of funds among various programs in our system of government. However, particularly since the successful flight of SpaceShipOne in mid-June 2004, which was financed entirely by private funds, the private sector will have an increasing influence on the investments in the future exploration of space.

It is worth examining the validity of arguments that we should only send robotic missions to the Moon or to Mars, and not human missions, because the latter are more costly and are a risk to human lives. I should like to submit to the reader of this article that, contrary to such contentions, our quest of scientific knowledge and our desire to explore the unknown come from the same basic drive that is present in all successful human groups that have survived the ordeals of nature—evolutionary selection for survival. We are referring here to the desire to understand the world in which we live—islands, valleys, mountain ranges, savannas, continents, planets, solar systems, and galaxies alike. The desire to understand nature through scientific quests and the spirit to explore and find out personally what's beyond the mountain range and the ocean are entirely compatible since they are the two sides of the same survival trait that has thus far made an evolutionary success of the human race.

There can be no question that both humans and robots are valuable in space exploration. Robots have proven their ability to revolutionize scientific understanding of the universe. Moreover, as Ranger and Surveyor did for Apollo, robotic spacecraft make excellent scouts for future human missions in the solar system. At the same time, humans can be integral to maintaining spacecraft health, as shuttle astronauts have done several times for the Hubble Space Telescope. And, in the future, it may be the cognition skills that human space explorers possess that will answer the questions, for example, of where, when, and how much water once existed on Mars.

I believe that a society that would want to send astronauts or cosmonauts into space would be one that would willingly support "pure" research in space as well. And indeed, this has been the case in the United States. Quite understandably, the voting public finds it easier to identify with astronauts and cosmonauts and their great adventures in space. On the other hand, they are thrilled at the photographs returned from space, sometimes with unanticipated consequences. Take, for example, the Apollo 8 photograph of the blue and fragile-looking Earth hanging over the lunar horizon; taken by astronauts, it was expressly this photograph and the sense it conveyed of a delicate spaceship Earth that gave a major impetus to the environmental movement. Images of astronomical objects transmitted from the Hubble Space Telescope have also excited the population at large and have inspired the younger generation to explore the depth of space.

The political sphere has been equally—and rightly—attentive to both human and robotic space exploration. During the peak of the Apollo missions, a great deal of money became available for building orbiting astronomical observatories and other space probes, too, although these spacecraft generally had little to do with the success of the Apollo missions. Today, manned missions in space, Mars rovers, and an assortment of astronomical and Earth observing spacecraft all are supported within NASA's budget.

The availability of economical and safe reusable launch vehicles is in clear sight now with the success of Burt Rutan's—and Paul Allen's—SpaceShipOne. The easy access to low Earth orbit is crucial in the future development of space. In terms of the energy requirement, reaching LEO is really getting half-way to anywhere in the solar system, as Robert A. Heinlein wrote so prophetically a few decades ago.

Indeed, human missions and robotic instruments soon will be working with each other to fulfill humankind's destiny of expanding our domain to the Moon and someday to domains beyond—even to distant solar systems. President Bush put forth a vision for the U.S. civil space program in which this very idea would be made reality. Within this vision, robotic explorers would conduct reconnaissance of the Moon, Mars, and beyond, to be followed by human explorers when the time was right. But even during human expeditions to locales throughout the solar system, robots would accompany people to support the missions as necessary. Human-robotic cooperation will take a multitude of forms, some of which we are not yet even aware. Perhaps after establishing a considerable human presence and permanent base on the Moon, we could launch both robotic and even human interplanetary missions, possibly using solar-powered magnetic catapults (so-called mass movers)—missions that are currently quite expensive if done directly from Earth. We could also use the Moon as a stable optical bench in space tens of kilometers across to conduct such important astronomical observations as broad baseline interferometry.

Finally, it is of importance to remind ourselves that the exploration of space—be it with manned missions or employing robotic instruments—is for the purpose of enabling and aiding the human exploration of space, physically and intellectually. The improvement of our knowledge of the universe through scientific undertakings is for the benefits of the human race, and not just for some abstract concept called 'pure' research.

Sooner or later, probably much sooner than many might think, we will need to reach out into space for various reasons involving our survival. For example, our fossil fuels, especially oil, will become more scarce and will eventually be exhausted in a matter of decades—not centuries. Solar power satellites (SPS) would, for example, offer an alternative—and virtually inexhaustible—source of energy. Even with the current state of technology, an SPS, in a geostationary orbit, with a collecting area of several square kilometers could continuously transmit to the rectenna (receiving antenna) on the ground about one gigawatt of electricity, which is roughly equivalent to the power generated by an atomic power station. Coincidentally, we might note that the environmentally benign sites for hydroelectric power plant dams have been virtually exhausted. [Let us by all means utilize the solar power on the ground as much as practical but that has its limitations due to the diurnal and seasonal cycles of the solar radiation; it also depends very much on the local weather as well.] Another example would be platinum; the availability of platinum, an essential element in the new, environmentally-benign energy technology of fuel cells is quite limited. Currently, it is mostly mined in a region of Africa where an earlier meteoroid impact left the platinum deposit. Mining asteroids—or looking for platinum deposits on the Moon left by meteoroid impacts there—might make economic and industrial sense in the case of such a precious metal. The list of potential benefits from space exploration can be endless.

The Sun is evolving as it exhausts its hydrogen fuel in its core. It will gradually evolve into a giant star. Somewhere along the line, as the Sun evolves, the Earth will cease to be hospitable as a habitat for the human race. We may need to terraform Mars, or some of Jovian and Saturnian moons, and migrate there. That will be a long time down the road. But unless we become a space-faring race when we can, that option may no longer be available. A race that does not explore its frontiers when it has the opportunity to do so may never have the energy and will to undertake it later, possibly thus dooming itself to extinction.

Too often, human space exploration and robotic space exploration proponents argue for their programs at the expense of the other. The "space pie" need not be divided by such differences. Both human and robotic space missions in space ought to be supported together. They are complimentary to each other. In an age, in which both undertakings are not only feasible but also pragmatic, neither would likely do well without the other—technologically, scientifically, financially, culturally and politically.

Yoji Kondo holds a Ph.D. in astrophysics from the University of Pennsylvania, headed the astrophysics laboratory at the Johnson Space Center (formerly, Manned Spacecraft Center) during the Apollo and Skylab Missions, and was the NASA director of a geosynchronous satellite observatory for 15 years, among other roles. He also served as President of the International Astronomical Union's (IAU's) Commission on "Astronomy from Space", as well as President of IAU Commission on "Close Binary Stars" and the IAU Division on "Variable Stars." He organized the IAU Symposium on "Examining the Big Bang" with Menas Kafatos, and co-edited and published its proceedings (Kluwer Academic Publishers). He has also edited or co-edited ten other scientific volumes since 1975, including "X-ray Binaries," "The Local Interstellar Medium," "Exploring the Universe with the IUE Satellite," "Evolutionary Processes in Interacting Binary Stars," "Observatories in Earth Orbit and Beyond," "The Realm of Interacting Binary Stars," and "Space Access and Utilization Beyond 2000." Among the professional recognitions Kondo received are the NASA Medal for Exceptional Scientific Achievement, the Federal Design Achievement Award (concurrently issued with the U.S. Presidential Award for Design Excellence), and the National Space Club Science Award—in addition to seven other awards from NASA, Johnson Space Center, Goddard Space Flight Center, and the European Space Agency. His avocation is judo (6th degree black-belt) and aikido (6th degree black belt), and has been teaching the martial arts for a few decades.

Alex Gimarc is a retired USAF Lt. Col. who lives in Anchorage, Alaska with his wife Diana and youngest son Grant . I first met Alex at a Space Studies Institute (SSI) conference in Princeton New Jersey. A low-key and friendly fellow with a southern accent and a real "aw shucks" attitude, his style belies a sharp and incisive mind, and a strong knowledge base, not just of space issues, but history as well. Not the image of a space "geek", Alex was running the External Tank Project for SSI, which he did for over a decade. He has been active in the space entrepreneurial community for over 20 years. A plain spoken man, he writes the same way. This essay offers Alex's common sense, no nonsense take on why a free enterprise approach to the frontier is the only way to make our breakout permanent.

History and Frontiers—What Works. What Doesn't.

by Alex Gimarc

Today we stand on the threshold of a very interesting and exciting century, providing we as Americans make the right choices. I will not argue whether or not humanity ought to go into space, for we have been doing so for over four decades, and I believe that we should continue to do so—just in much larger numbers. The next questions and answers now become far more important: How do you go there ? Who pays for it? And how do you structure your efforts so this happens quickly, efficiently, safely and affordably? All three of these questions must be answered within the framework of getting large numbers of people permanently off this planet.

The entire discussion must be oriented toward how to build a new economy off planet. American taxpayers simply will not—and should not be asked to—ante up the hundreds of billions of dollars necessary to fund a new and "exciting" all NASA effort to build a base on the Moon and fund a trip to Mars. The taxpayers will not—and should not be asked to—fund a military effort to do the same thing, for there is no alien threat (that we know of yet) out there. Neither approach will lead to large numbers of people permanently living and working off planet.

On the other hand, American entrepreneurs are already perfectly willing to expend a huge amount of time, money, and creative effort creating new products, new services, new jobs, new infrastructure and most importantly, new wealth in their drive to create profit-making businesses off planet. The powerful engine of a competitive free market and the subsequent creation of vast amounts of new wealth will fund the move of humanity off planet. It is the only thing that will get the job done.

Why is this important? It is important because the free marketplace—the actions of millions of our neighbors making decisions in their best interest—is the mighty engine that will power our move into space. The free market transmits information about those decisions incredibly fast and efficiently, so that other personal and economic decisions can be made. The actions of companies in a free marketplace also creates new wealth, new jobs, and grows economic activity. People and businesses participating in free and open competition for customers are the best single way of creating wealth. And if we are to permanently leave this planet, someone other than the taxpayers will have to pay for it.

Before we discuss why a market-based space effort is absolutely crucial, I need to make a few comments about free-market basics to set the stage for the ongoing discussion.

The best description of what you need for a free, open and competitive marketplace comes out of the Chicago School economists, most notably Nobel Laureate Milton Friedman. Students can follow these ideas back through the Austrian School with Friedrich Hayek, and then back to "The Wealth of Nations" by Adam Smith. The principles are pretty simple in the abstract, and pretty difficult to adhere to in practice. What do you need to have a free market? You need a free, open and competitive marketplace, property rights, a relatively stable legal system, minimal regulations, all underlain by a basic morality (religion).

That's it. That's all it takes.

Economic success is measured on a sliding scale. The more free your marketplace, the better your economy performs. The more stable your legal environment, the better reward your risk-takers get for their actions, and in turn, more risk is taken and more innovation happens throughout the economyAlso, the more benign and unobtrusive your regulatory environment, the better your economy performs. The tool used by the fascists (National Socialists) was to allow private property to be held, but to completely regulate its use. Also think of "public-private partnerships" as yet another example of what not to do. The more solid your protection of private property rights, the better reward your risk takers get for their efforts, and better performance of economic activities overall. Finally, the entire enterprise needs to be based on a moral foundation to guide business decisions.

History: Great Leaps vs. Incremental Steps (Follow the Money)

Humanity has entered a number of new frontiers over the last several centuries. There are two basic techniques—the Great Leap Forward, and the Incremental approach. Normally, a combination of both is used, with the Great Leap Forward coming first (think the Apollo program's flag and footprints missions to the Moon). The nation that can most quickly exploit those visits with permanent presence ends up owning the new frontier.

The move into a new frontier is normally done in incremental steps, with the marketplace making the decisions about what sort of things, what sort of skills and experience are necessary to support and exploit that move. Here are a few examples of how to do the move into a new frontier correctly and how to not do it correctly.

New World – 15th – 17th Centuries

The exploration of the New World half a millennia ago was not triggered as a result of the Spanish decision to fund Columbus' "missions" into to the new world. There was a tremendous amount of exploration—incremental steps—done before the decision to fund Columbus' "missions."

The world's best navigators at the time were the Portuguese. They were slowly, and at times not so slowly, working their way via multiple missions down the west coast of Africa. Each trip built upon the one before, refining navigation techniques, improving maps, discovering new lands, trade items and people. Daniel Boorstien in "The Discoverers" described this incremental approach well.

The economic engine driving this interest was an effort to discover an easier trade route with the East (China and India) than the overland route used by Marco Polo in the 13th century. There was a growing trade in products that could not be obtained locally. During this growing trade, the already robust network of sea and river ports around Europe and the throughout the Mediterranean was improved to support the newer, larger, and faster ships necessary for the longer exploration trips.

Once the western hemisphere was discovered, another round of incremental exploration began. Nations that allowed some level of private and commercial ownership retained their new colonies, created new wealth, and grew long-term presence in the Americas. It is no accident that the Spanish, which came with Viceroys (Vice Kings), soldiers to conquer and plunder, and priests to convert the savages, did not have enough return of wealth from the new colonies to sustain a permanent presence in the Americas. When the gold ran out, they had to turn to other products and had not set up a system to create new wealth. The British, on the other hand came with shopkeepers, some level of property rights, and had their permanent presence powered by a variety of companies. The Brits stayed, and their presence grew, because even at the dawning of capitalism, their economic methods of exploring and creating new wealth in the New World were just enough better than the Spanish that they won the economic competition and dominated the world economically for centuries afterwards.

Space policy expert Jim Muncy notes that the Spanish approach was successful early and quickly returned a quantity of gold and other riches to Spain. The somewhat more market based English approach, started 120 years after the Spanish, but was far more successful and sustainable in the long term. 500 years later, people travel North rather than South when they emigrate in the Western Hemisphere.

The other thing that happened over a period of time was a growing resistance of the general population, Parliament in England, and opposition among other governing bodies and individuals in Spain and other nations to continue to pay for development of the colonies overseas. When the incoming wealth decreased quickly, like it did for the Spanish, public, national and royal interest also went away. It took the British a while longer to lose interest, because they took the time to actually build a relatively non-government funded economy. However, by the time of the US Revolution, there was a not insignificant backlash against the continuing economic drain on the British economy, despite the fact that the colonies were taxed and viewed as a revenue source at the time.

The great exploration voyages of the Fourteenth and Fifteenth centuries would have never happened without the seafaring infrastructure already in place, the skills already learned, the ports already built, and the trade routes already known. Humanity over the course of thousands of years learned how to sail the Mediterranean, learned how to trade items of value, learned how to build ports, learned the skills necessary to build, support and man the ships. Over time, they expanded those trade routes, that network of ports, skills and training around all of Europe. Eventually that network was expanded around the Horn of Africa and across the Atlantic.

If we take this analogy forward today, take a look around and ask how can we possibly mount and sustain a colonization effort to Mars without having a significant private and commercial presence on the Moon? How can we have a significant permanent presence on the Moon without having one in Low Earth Orbit or throughout CisLunar space? The answer is that we simply can't. There is not enough money. Someone other than the taxpayer—the businessmen, entrepreneurs, customers and their stockholders—needs to pay for that permanent presence through a series of personal economic decisions. We are indeed fortunate that there was no "New World Program" or "National Seafaring Administration" with a monopoly on sea travel in place during the 14th and 15th Centuries.

To be fair, at the time there was a command-driven exploration program out of Spain which used the Spanish Navy and Army. The British on the other hand mixed it up a bit, with the Royal Navy and the Merchant Marine. This difference in approach was enough to make all the difference.

American West – 19th Century

The expansion of the United States across the North American continent was yet another example of the use of private property, and a free and relatively open marketplace to guarantee that expansion. Land was acquired via treaty and war. It was then turned over to territorial and state governments to dispose of. Congress also set up a number of land giveaways to the citizens and corporations via a series of Homestead Acts throughout the 19th Century and a series of Pacific Railroad Acts from 1862 – 1874. A system of public Universities was created via the Morrill Land Grant Act of 1862.

The common thread here was that Congress chose to transfer property from government ownership permanently to the citizens. The government got out of the way, except for providing for the common defense, and allowed the economic expansion of the United States across the entire continent.

Congress also took the lead in fostering development of infrastructure across the American West, in this case the transcontinental railroad system. They did it via a series of Pacific Railroad Acts, which conferred property rights along the railroad routes. They then got out mostly way and allowed the marketplace to operate.

Alaska – 20th Century

The United States purchased Alaska from Russia in 1867 for around 2.5 cents per acre. It was administered as a territory until the Alaska Statehood Act of 1958 was passed by Congress. Part of the local push for statehood was in response to federal mismanagement of fish and wildlife (a property rights dispute). Alaska entered the Union with over 60% of land held by the federal government. Most of the rest is held by the state itself, with over 40 million acres conferred to Alaska Natives via two Congressional Acts in 1971 and 1980. Less than one percent of land in the state is privately held. Mineral rights are commonly held, which is normally a mistake, as it will tend to limit investment, exploration, ownership and competition.

Mineral exploration on state lands is conducted under the auspices of a series of state leases, with new lease sales being offered up by the State of Alaska every year. One early series of lease sales led to the creation of the Prudhoe Bay oil fields, a source of great wealth to citizens of the state since the 1970s. There is currently a competition underway to privately fund and build a natural gas pipeline from the Prudhoe Bay oil fields.

Economic mistakes made in Alaska include the huge percentage of federally owned and administered land, common ownership and management of mineral rights, common ownership and management of fisheries and game, and the miniscule amount of private property held statewide. Yet with all of these mistakes, there is sufficient economic activity, innovation, and competition to construct a self-sustaining economy with only 600,000 people in the state.

If we can colonize and lead productive lives in an area that regularly gets -40-50 degrees below zero, we can probably figure out how to do it in orbit, on the surface of the Moon, on the asteroids, on the comets, on the Martian moons, or on Mars itself. But in every case, you must grow that economy incrementally in a competitive, open, free market <u>Antarctica – 20th Century</u>

Antarctica is yet another example of incremental steps in exploration and building infrastructure. It is also an example of what happens to an economic expansion if you get in the way of economic basics.

The exploration of the continent during the 19th and early 20th centuries is the stuff of legend. Antarctica is very difficult place to live and work. You can die in a matter of minutes if you go outside unprotected at the wrong times of the year. Yet we see people working year round and making money on the Alaska North Slope, the Canadian Arctic, Siberia, and other Northern locales. Why is Antarctica different ?

Antarctica is a pretty big place, mostly—but not all—ice covered. There are minerals, open, unclaimed stretches of land, and spectacular places to see and experience. You would think that there would be permanent commercial human presence, hotels and related tourist complexes. Yet there aren't. Antarctica's ownership is controlled by international treaty, the most recent being the Antarctica Treaty of 1959, which designated the entire continent as common heritage among the signatories. This essentially eliminated all private property rights, locked up an entire continent 75% the size of the Untied States, and turned it into the private playground for scientists. Not surprisingly, the scientists vigorously guard their taxpayer-funded and supported playground.

Today there is a small, but flourishing tourist trade operating around the edges of the continent, but everything brought in must be removed when they leave. Adventurers are even denied assistance, as happened to an Australian individual attempting to fly across the South Pole in late 2003. He miscalculated the fuel, and had to land early. Scientists at the American McMurdo Base and the New Zealand Scott Station refused to sell him fuel, saying they weren't running gas stations. The scientists are resisting new tourist travel on the continent in order to defend their turf.

Today, Antarctica is an example of what happens when property rights are denied and a government monopoly—this one scientific in nature—is created. Rather than being a new job and wealth creator, activities on the continent are net expenditures to the taxpayers of the signatory nations. There is no growing infrastructure in and around the continent. There is no self-sustaining economy. There are few happy and satisfied customers for goods and services who except for a few ongoing tourists that visit for a bit and then have to leave.

There are those that have tried over the course of the last 40 years to turn the rest of the Solar System into the "common heritage of all mankind." Places that are managed nationally or internationally as "common heritage" never allow the creation of private property or the free markets essential to guarantee human expansion to those places. Turning the Moon, the asteroids, comets, Mars and its moons into "common heritage" would be the single best way to guarantee humanity never leaves this planet.

In this regard, Antarctica becomes yet another example of how not to put people permanently in a new location.

What Works ?

The preceding examples illustrate the differences between successful and failed approaches to exploration and expanding human presence into a new place. When governments treat a new frontier as an opportunity for their citizens, confer basic property rights, get out of the way and allow a marketplace to take root and grow, economic activity will take place. People will pursue their dreams of wealth, liberty and simply their desire to go somewhere new, see something new, and do something new.

Contrast this with recent examples of how not to permanently colonize a new frontier and create a permanent presence in space. As boomers, our parents were asked to support Apollo, because it was a way to demonstrate our superiority in competition with the Soviets. At the time, this was a relatively easy sell, for it happened at the height of the Cold War, when intercontinental ballistic missiles (ICBMs) were the weapon of choice. We set up a Soviet-style, centrally planned and executed program, executed it superbly, and beat the Soviets to the Moon in 1968 and 1969. Once we won the competition, like most sports fans, we went on to the next event—which did not include return trips to the Moon. We treated the Moon as a one-shot activity, with the goal of simply beating the Soviets to be the first to plant a flag. We were wildly successful. We put 12 men on the Moon. Yet it has been 32 years—an entire generation—since we've been there.

In the 1970s the taxpayers were asked to build a cheap, reusable, affordable way to go to and from space. Congress grudgingly went along, but did not provide a blank check this time around when they approved the Space Shuttle. The Nixon, Ford, and Carter Administrations, Congress, and NASA all decided that the way to reach this goal was once again a centrally planned, centrally executed Soviet-style program. No alternatives were seriously considered which would foster competing approaches including, private or commercial ownership. No property rights were conferred. No free, open and competitive marketplace was allowed to develop. Even the Department of Defense (DoD), which had a real, demonstrated need for placing a variety of objects into orbit on short notice, was forced to become part of the monopolistic Shuttle program.

Only five shuttles were built—four initially and one to replace *Challenger*—all part of a fleet that was at the time touted to someday be flying once a week. Sadly, when you have a lack of competitive marketplace, a lack of private ownership, and a government monopoly so tight that even inter-departmental competition is halted, you tend to make bad strategic design choices, which become harder to rectify as time passes. By the time the shuttle started flying in the early 1980s, the promise of easy access into orbit for the government, let alone all Americans—was shown to be an empty promise.

Finally, we have the case of the International Space Station (ISS), which went from a facility supposedly able to house a crew of 8-12, constructed over the course of 6-8 years for less than $10 billion turn to an economic disaster That is still not finished at the time of this writing. By the time ISS is "completed" in the next decade, we will have spent well in excess of $100 billion to construct a government owned and operated facility capable of housing a crew of 2-6 maximum. No commercial entity would allow such an economic disaster to take place—in fact it would probably go bankrupt and/or its stockholders would fire the management—if they weren't indicted. The unfortunate part of this story is that a space station is essentially a free-flying pressure vessel. Pressure vessels are pretty easy. We have been building them for generations for aircraft, submarines, boilers, chemical and petroleum applications. Yet, for some reason, under government management, and as a legacy of NASA's "space is different and only we can do it" legacy, we made it hard.

Congressional support for ISS was not particularly easy to get, and it shouldn't be, given the economic mess the ISS has turned into.

We have seen over time, that American space efforts have become increasingly difficult to sell to the general public. They have required significant expenditure of political capital, at a time where such expenditures are better focused elsewhere. There is no way that Congress will fund a manned trip to Mars, especially if we are to do it right, constructing support infrastructure in Low Earth Orbit, on the Moon, on the Martian moons, and a permanent base on Mars. The President's so called "Vision for Space Exploration" (VSE) notwithstanding, I believe that in the end all that will be funded, if anything, is a single flight to plant a flag, take a few photos, and return a few rocks. It will all be done at taxpayer expense. It will not confer any ownership or participation on any private citizen. And most importantly, we will only do it once.

Discussion

A free market is vital, for it is the mechanism by which customers and vendors create and refine new goods and services incrementally in a competitive environment over a period of time. For example, several hotel corporations, all competing for high-end customers would never have turned a 6-8 year, $8 billion facility like the International Space Station into a 15 year, $100 billion public works project. They would have figured out how to launch a turnkey platform, a space hotel, capable of serving 50 people, for less than $1 billion, in a single flight. They would have done it quicker, more robustly, less expensively, and figured out in very short order how to use such a facility to create wealth, revenue and increase both their customer satisfaction and stock prices. If a single orbiting hotel was not sufficient for their needs, they would have launched another 2, 3 or 20.

If you then get several companies competing for space tourist dollars, you force all costs down, and you create a market for a robust manned launch capability. If we are to return to the Moon, we need both. If we are to go to Mars, we need something to force the costs of living, working, and traveling in space down. We need something to make space operations inexpensive, safe, sanitary, and profitable. We will not get this with a Mars Direct approach, for it builds no infrastructure. We will get it with a market-based return to the Moon and then on to the asteroids—and yes, Mars.

How to Return to the Moon

Can we go back to the Moon as a first step ? Absolutely. We did it once before, from a standing start, knowing absolutely nothing about living and working in space, and were standing next to our lander on the surface of the Moon within 8 years. We don't need to wait until 2020, nearly 60 years after Apollo 11 to return. However, if we don't do it the right way—on the shoulders of entrepreneurs, risk takers, businessmen and wealth producing Americans, it will indeed take at least that long, if we end up doing it at all.

If we want to get off this planet permanently, there is no safer, more affordable, more robust way to get people into space than via an open, free marketplace. The steps from here to wherever we want to go will be incremental, unplanned, marketplace-driven choices by businessmen, visionaries, board members, customers and stockholders. We will go to stay, permanently, and start making money, raising children and grandchildren in and on the New Worlds.

Alex Gimarc earned a BS in Aerospace Engineering and a MS in Space Technology. He authored the SSI Report on External Tanks presented to the Reagan National Commission on Space in 1986. Alex flew fighters while in the USAF and worked operational requirements for military space systems at Air Force Space Command. He has published numerous space and related papers over the years; and has spoken extensively on the subject.

In 2004, I testified on the future of NASA before the Senate Space Subcommittee. Along with me sat Dr. Howard McCurdy of American University, who, although coming from a much more traditional approach to space exploration, shared my disdain for the directionless NASA culture that has characterized our space program for decades. In his testimony, he pointed out that, if managed correctly, the push to the Moon and Mars could result in a much more focused and efficient exploration agency. Dr. McCurdy's orientation is far less free-enterprise than my own, and many of the other essays in this book, but his point regarding the internal structure of our space agency, and the need for clear, concise and well-defined goals as a means to bring out the excellence we must have if we are to succeed, is completely valid. If we can transform NASA into the "lean mean exploration machine" I mentioned in the introduction, with operational activities managed by the private sector, we might have a chance at success as we Return To The Moon. Thus, when this book was being planned, I asked Dr. McCurdy to contribute this updated version of his Senate testimony.

Returning to the Moon Will Transform NASA

by Howard E. McCurdy

Advocates of space flight offer many reasons for returning to the Moon and exploring Mars. Extensive investigation of nearby spheres will allow scientists to better understand the processes that make Earth livable. Exploration inspires human creativity and invention, encourages science and technical education, and promotes international cooperation. It gives humans the capability to move themselves and their machines around the inner solar system. It satisfies the human need to migrate and may someday diversify the species onto more than one planet.

These are noble purposes, ones sure to affect human civilization across very long periods of time. The objectives set forth in the president's Vision for Space Exploration would have a more immediate benefit as well. They would transform the National Aeronautics and Space Administration (NASA), the government organization set up nearly fifty years ago to lead the national civil space endeavor. One of the principal reasons for going back to the Moon and onto Mars is the certainty that the mission will transform NASA.

Thirty years of drift in the human space flight program have left NASA ill-prepared to undertake a focused exploration program, especially one that respects the cost constraints imposed by the president's directive. Yet this need not cause despair. The people who work on the national space effort have overcome similar difficulties in the past and they can do so again. The agency created in 1958 was not capable of sending humans to the Moon, but eight years after President John F. Kennedy issued his famous declaration Americans stood on the surface of another world. In the same way that the lunar objective proved to be a transforming event for NASA in the mid-twentieth century, the Moon-Mars objective can transform the agency in the twenty-first. The fact that the agency has transformed itself in the past encourages us to believe that this can happen again.

Why NASA is Not Prepared

For more than thirty years, successive leaders in the field of space exploration have called upon public officials to give NASA purpose and direction. As a science and engineering organization, NASA works best when its workers possess a clear vision of their ultimate objective, their schedule for accomplishing those objectives, and the various cost and risk constraints under which they must operate.

From 1961 through the landings on the Moon, the human space flight program operated under such mandates. The purpose and timetable established in President John F. Kennedy's May 25, 1961, speech provided focus for America's civil space effort and imposed discipline on the new space agency. People working on the national space effort sought to land humans on the Moon and return them safely to Earth by the end of 1969 at a cost then estimated to lie between $19.5 and $22.7 billion. The mission was accomplished in July, 1969, at a cost (through the first landing on the Moon) of $21.4 billion.

As Americans prepared for the lunar landings, NASA officials and other public leaders contemplated the steps that lay beyond. Anticipating the current presidential initiative, they called for a post-Apollo space effort concentrated on the Moon and Mars, bolstered by an energetic space science program. The report of the Space Task Group, presented in September, 1969, outlined an objective-rich effort of human and robotic flight. In March, 1970, President Richard Nixon rejected the report of the Space Task Group, thereby initiating three decades of drift in which the leaders of NASA's human space flight program operated without long-term focus and direction.

In response, leaders of the space community adopted an incremental approach to human flight. They pursued elements of their long-range vision in succession, one at a time, without reference to any overarching goal. In governmental policy-making, incremental policies have a strange effect, a consequence of the willingness of public leaders to approve individual projects without agreeing on long-term objectives. Because long-term goals are not specified, public leaders can agree to support individual projects for different and often conflicting reasons.

Where incremental decisions collide with the harsh realities of physics, the results can be disastrous. Without long-term goals to provide purpose for large-scale engineering endeavors, the people promoting such projects typically create broad political coalitions as a means of getting new initiatives approved. The projects emerging from these coalitions often contain conflicting objectives, so many that the projects are physically difficult to carry out.

The people designing NASA's space shuttle were told to build a vehicle that could carry people to and from an Earth orbiting space station; transport the components of that station to space; serve as a "space truck" for commercial payloads (some carrying upper stage rockets attached to payloads headed for geosynchronous orbit); deliver military reconnaissance satellites; deliver, repair and possibly return space telescopes; and serve as a short-duration microgravity research laboratory. The whole shuttle fleet had to be developed for $8.05 billion, fly for about $14 million per mission, and operate between 24 and 50 times per year. According to the original cost and schedule estimates, shuttle managers were to have begun routine flights in 1979 and retired the fleet of five orbiters in about 1990. The total program was to have cost $16.15 billion, a substantial savings over the price of expendable launch vehicles at that time. All of those costs were stated in 1972 dollars.

Since NASA concluded the space shuttle test flights in 1982, it has flown the shuttle an average of five times per year. Per mission, operational costs have run eight times the original estimate—about $560 million per flight in inflation adjusted 2002 year dollars. Converted to 1972 dollars, that amounts to $118 million per flight—well above the $14 million estimate. Throughout the entire flight period, program managers have spent funds on shuttle capabilities and upgrades never anticipated in the original cost estimates. Summing the capital investments, operational activities, upgrades, and facilities construction, and adjusting for inflation,

NASA by the end of 2004 will have spent as much money building and operating the space shuttle as it paid to put the first Americans on the Moon.[1]

Regardless of what one thinks about the technical capabilities of the space shuttle (as a reusable space plane with wings, it accomplished many of its astronautical goals), the program has fallen far short of its cost, schedule, and mission objectives. Moreover, it contributed to an organizational culture of normalized deviance that weakened NASA's human space flight effort.

Throughout the shuttle era, NASA employees and their contractors came to accept known flight risks as the normal way of doing business, substantially eroding the high performance practices that NASA's founders created in the first decade of space flight. These alterations are well documented in the reports of the commissions that investigated the space shuttle Challenger accident and the loss of the space shuttle Columbia. As an illustration of the manner in which NASA's human space flight culture changed, consider the follow point. NASA officials and their contractors in the early decades of space flight operated under the assumption that the agency would not launch a spacecraft until its designers could prove that they were ready to fly. In both the Challenger and Columbia accidents, NASA officials required concerned individuals to prove that spacecraft were not ready to fly (or land) in spite of visible safety concerns. Cultural beliefs, a powerful force in shaping institutional behavior, most commonly consist of the assumptions that people make as they go about their work.

Incremental space policies contributed to this effect. As members of the Columbia accident investigation board observed, the existence of so many conflicting objectives within the shuttle program severely compromised NASA's ability to build a safe and reliable vehicle. The extensive mission requirements, themselves burdensome and contradictory, competed with cost and schedule goals. Substantial capital investments were needed to produce a vehicle that cost less to fly and compressed flight preparation times. But political leaders wanted a shuttle with low capital development costs too. "The increased complexity of a Shuttle designed to be all things to all people," board members wrote, "created inherently greater risks than if more realistic technical goals had been set at the start." The most serious mistake that NASA officials made in developing the vehicle dealt not with the design of any particular component, "but rather with the premise of the vehicle itself." (CAIB report, p. 23)

NASA officials undertook a similar approach with the design of the International Space Station. They appealed to astronomers, people interested in space science, advocates of a return to the Moon, commercial interests hoping to manufacture micro-gravity products, communication satellite companies, international partners, and the U.S. military. The early space station design included hangers for satellite repair, micro-gravity research laboratories, mounts for observational instruments, pallets for scientific instruments, and two large keels within which large spacecraft bound for deep space missions could be prepared. Further confounding these objectives, NASA officials estimated that they could develop such a multi-functional facility for only $8.8 billion and complete its assembly in ten years.

The people at NASA headquarters who conceived the space station concept believed that agency employees could meet these cost and schedule constraints. They sought to do so through a judicious combination of technology improvements and severe limits on the tendency of field centers to impose expenses not directly related to the fabrication of hardware. Even allowing that an internal Concept Development Group set the estimated cost of the space station closer to $12 billion (in 1984 dollars), NASA officials fully failed to meet their cost and schedule goals. The actual space station, including design, hardware, operations, research, technology, and transportation, will cost more than $50 billion.

Space station advocates learned that the political coalitions necessary to win approval for the orbiting facility were much easier to construct than the actual facility. While attractive for building political support, the various activities planned for the space station proved technically incompatible and impossible to develop within the proposed cost. As a consequence, NASA officials spent the entire ten years set for construction of the station (1984-1994), as well as the original $8.8 billion cost estimate, redesigning the facility and reducing its scope. When redesign work ended in 1993, NASA officials wrote off the whole $10.2 billion spent that far. A new cost estimate of $17.4 billion then grew to $25 billion. That sum, however, did not include the cost of transporting the station to orbit, the cost of operating the station for ten years, nor the money spent during the redesign years.

Program delays accompanied cost growth. Originally set for completion in 1995, the date for finishing the U.S. station core moved to 2004 and beyond. Even with the shuttles operational after the post-Columbia stand-down, the station will require several more flights just to reach U.S. core complete and a grand total of twenty-five shuttle launches and one Russian Soyuz to conclude assembly.

The inability of space station managers to meet the original completion schedule retarded pursuit of initiatives that lay beyond. Within the framework of incremental policy making, NASA officials were obliged to pursue new human flight initiatives one at a time. The incremental approach contributed to a mentality within NASA in which new ideas sat on the base of old endeavors. This was dramatically demonstrated by NASA's reaction to the 1989 Space Exploration Initiative.

As dissatisfaction with the unfocused approach to human space flight grew, a variety of people urged national leaders to abandon incremental policies and reinstitute a long-range space goal. Yet NASA officials had grown accustomed to the practices necessary to operate in an objective-free atmosphere. The effect of this mentality was readily apparent in the agency's response to President George H. W. Bush's 1989 proposal for a human Space Exploration Initiative focused on the Moon and Mars. Following the proposal, White House officials directed NASA employees to prepare an enabling plan. NASA officials treated the Space Exploration Initiative as a healing balm, an ointment applied to institutional members to make them well. NASA's so-called "90 Day Study" called upon Congress and the White House to expand the space shuttle fleet and enlarge the international space station, programs already underway. The interplanetary spacecraft proposed by agency planners and based on conventional technologies proved so large and costly that one commentator labeled it a "death star." Excessive hardware along with official cost estimates exceeding $500 billion killed the initiative before it could begin. To people outside NASA, the agency response seemed more like an exercise to protect existing projects and restore the health of ailing field centers than an opportunity to pursue a long-term vision in space.

People within NASA and its contractor workforce commonly treat cost, schedule, and reliability objectives as factors to be traded one against the other, in which the gains in one or another must be compensated by losses in the last. This "pick two" philosophy is well imbedded within the current human space flight culture. Yet leaders in American industries that produce items such as computers and micro-electronics commonly expect their workers to simultaneously improve cost, schedule, and reliability goals. During the 1960s, NASA officials and their contractors met severe cost and schedule constraints on Project Apollo without sacrificing overall safety. NASA project managers took substantial, calculated risks to do so, such as the practice of "all up" testing of the Moon rocket and the decision to send astronauts on a circumlunar voyage on a flight immediately after one on which the Saturn V rocket engines malfunctioned. In spite of this heritage, workers in the civil space agency, at least on the human space flight side, strongly resisted pressures to simultaneously improve cost, schedule, and reliability during the period of incremental drift.

During the quarter century that followed the landings on the Moon, NASA's human space flight program underwent changes that compromised the ability of government workers and their contractors to carry out complex human space flight activities in reliable ways. The agency went from an institution capable of meeting multiple goals to one in which human space flight officials struggled to achieve the objectives imposed upon them.

Organizational practices take root over many decades. Similarly, reforms require many years to become imbedded in the minds and habits of agency employees. Organization cultures deteriorate slowly and they revive themselves only after lengthy adjustment periods.

Many factors promote organizational revival. New leadership, workforce turnover, the hiring of new employees, the creation of organizational symbols and stories, and extensive training all serve to encourage organizational revival when used in appropriate ways. Nothing encourages revival more than new challenges and objectives. In that sense, the Moon-Mars initiative has created an opportunity for the transformation of a NASA institution grown weak over the past thirty years.

How NASA Transformed Itself in the Past

The most important institutional transformations necessary to dispatch Americans to the Moon, when they occurred in the 1960s, came from sources outside the newly created National Aeronautics and Space Administration. NASA executives imported systems management techniques from the American military and rocket skills from the Army Ballistic Missile Agency. Systems management was developed during the 1950s as a means to push through the development and production of America's intercontinental ballistic missile system, a crash program as ambitious as the race to the Moon. Systems management techniques, which space historian Stephen Johnson has called "the secret of Apollo," were laid over an extensive reservoir of in-house technical capability contained in the government installations assembled to create the new civil space agency. The combination of systems management techniques and in-house technical capability, along with the necessary technology, placed Americans on the Moon.

In the spring of 1961, when President Kennedy challenged Americans to race to the Moon, NASA was institutionally unprepared to carry out the mandate. People from the agencies out of which NASA was formed three years earlier had great technical skill, but absolutely no experience managing activities on the scale of Project Apollo. They were accustomed to managing small research projects, not large-scale undertakings.

NASA officials at that time did not know how to manage large programs. Existing field centers were independent and uncooperative. The United States lacked the technology to fly to the Moon. No American astronaut had ever flown in orbit, much less engaged in rendezvous and docking. Leading strategies for reaching the Moon were either technically infeasible or impossible to complete by the decade's end. When NASA's head of human space flight made the suggestion that the agency concentrate all of its resources on solving the lunar conundrum, he was fired for what traditional interests within the agency viewed as intemperate remarks.

Yet eight years later Americans safely traveled to and returned from the Moon. During those eight years, NASA executives reorganized the agency twice, imported management experts from the ICBM program, and forced field center directors to report to coordinating program officers in Washington, D.C. They revised institutional procedures after the loss of three astronauts in a space capsule fire during a launch center ground test, a critical exercise in institutional learning. They invented new technologies and built new rockets. The person who oversaw NASA during the transitional period was an expert in management and finance. Neither an astronaut nor an engineer, James Webb had served as director of the U.S. budget bureau and later President of the American Society for Public Administration.

President Kennedy's May, 1961, speech proved to be a transforming event for the new space agency. It changed NASA from an agency of technical experts into an institution capable of implementing extraordinarily complex space flight activities. The lessons learned through the Moon race spilled over onto the space science side, where other project leaders carried out the great planetary and space telescope missions that followed.

Transforming NASA Again

To locate many of the institutional practices necessary to transform the civil space program in the 1960s, agency leaders looked outside NASA. They brought in people like George Mueller and General Samuel Phillips, experts with extensive project management experience but little knowledge of space flight. They imported large scale systems management. While some of the organizational reforms necessary to transform NASA in the twenty-first century will come from external business and government practices, many others already reside within the civil space agency. The institutional tools necessary to transform NASA exist within NASA, in old traditions and pockets of change.

Cost discipline: Successful completion of the Moon-Mars initiative will certainly require the restoration of cost discipline within the human space flight program. Cost and schedule discipline has been missing from the human flight program for thirty years, and the excessive cost expectations advanced by NASA officials during the 1989 Space Exploration Initiative sunk that early endeavor. The willingness of political leaders to provide the funds for the new space exploration vision depends considerably upon the restoration of a climate of confidence in NASA's ability to meet cost goals in the human space flight program.

Cost discipline did exist during the Moon race, when agency leaders struggled under substantial cost and schedule constraints. The notion that NASA executives met their schedule and performance goals because Congress relaxed the cost constraints for the Moon race is a myth. NASA employees struggled to meet the "end of the decade" deadline and maintain reliability within the $20 to $23 billion cost goal; agency executives fought hard to maintain sufficient funding, both in the White House and Congress. The largely invisible budget battles so exhausted James Webb that he resigned as NASA Administrator before the actual landing on the Moon.

Congress will not provide NASA with a blank check to pursue the current Moon-Mars initiative, no more than it did so for the first expeditions to the Moon. For knowledge about how to meet cost and schedule goals, NASA executives need look no further than their own space science activities.

During the past dozen years, NASA officials transformed their robotic space science programs. Space scientists significantly reduced the time and money needed to carry out robotic and satellite missions, often meeting severe cost and schedule goals. Consider this example of the transformations occurring in space science. Stated in the inflation adjusted value of 2002 dollars, the 1976 Viking mission to Mars cost over $4 billion. For that sum, NASA managers and their contractors successfully placed two Landers on the surface of the planet, and dispatched two orbiters to assist in exploration activities. The Mars Exploration Rovers that landed in January, 2004, carried out their missions for $820 million. Even acknowledging the added capability provided by the two Viking orbiters (a total cost in today's dollars approaching $1 billion), the difference is dramatic. NASA space scientists have learned how to fly for a fraction of the cost of previous endeavors, using robotic technologies that have advanced the state of space science enormously.

The current Moon-Mars initiative contains a similar challenge. NASA employees are being asked to return to the Moon, continue the current robotic exploration of Mars, construct telescopes that can inspect planets around nearby stars, and develop a new crew exploration vehicle—for no more than the agency spent to send the first humans to the Moon. The full cost of the Moon race through Apollo 11 totaled $21.4 billion, about $150 billion in the value of 2003 dollars. Extended over sixteen years, the financial allotment for new space exploration vision (in 2003 dollars) is $154 billion.

Compared to the shuttle-station experience, meeting this goal will require considerable cost discipline. On its face, the task may seem impossible. Nonetheless, NASA officials have encountered similar challenges in the past and prevailed. They met analogous standards in the 1960s and they produced low-cost innovation in recent robotics and space science programs. In the process, they have taken considerable risks and made occasional errors. In 1999, for example, NASA space scientists lost all of the low-cost spacecraft that they dispatched to Mars. Yet they learned from the experience and recovered, as did the people who struggled through similar lunar initiatives such as the failure-prone Project Ranger three decades earlier.

Project managers have achieved cost discipline in the NASA space science program through a number of related developments. Space science mission objectives have been focused and clear. Meeting cost targets has been elevated in importance to a level commensurate with science goals. Technology advances have produced smaller and more reliable spacecraft. New management techniques have been perfected, some significantly different than the large-scale systems management practices that propelled the success (and the cost) of Project Apollo.

Mission focus: Internal discipline typically requires mission clarity, a primary means for overcoming disagreements about agency priorities and the means for conducting space ventures. When planning for Project Apollo got underway in 1961, NASA employees had strongly diverging views about how to carry out the mission. Some wanted to build orbiting space stations or suggested that elements of the expedition gather at an orbital staging point above the Earth. Others wanted to build enormous rockets, and recommended a strategy called direct ascent. Not surprisingly, participants tended to favor those approaches that maximized the contribution of their own field centers. Different field centers wanted to be involved in different ways. The initial result was disagreement and deadlock.

An engineer from NASA's Langley Research Center tried to explain that Americans could not reach the Moon by the end of the decade unless they utilized a spacecraft that remained in lunar orbit while two astronauts piloted another vehicle to the surface of the Moon. At first, the idea of lunar orbit rendezvous seemed preposterous. The United States had not conducted a successful rendezvous in Earth orbit, much less one around the Moon. More significantly, the idea upset the plans of people with different agendas. The engineer persisted. "Do we want to get to the Moon or not," he asked. The question silenced critics. The discipline of the mission forced people to forgo vested interests and work toward their common goal.

For more than thirty years, people in the human space flight community have pressed government leaders to provide an overarching objective similar to the lunar goal. In their own history, NASA officials know how to respond to overarching goals. Their own experience with Project Apollo, as well as the more tightly focused space science programs, has shown NASA employees how mission focus can create internal discipline.

Management techniques: Return to the Moon and preparation for voyages to Mars within the envisioned cost constraints will require new management techniques. The systems management approach that NASA officials adopted in the 1960s provided substantial reliability, along with cost and schedule control. It also proved to be labor intensive and expensive.

Again, NASA officials can find examples of less expensive and more adaptive management techniques within their own agency. During the 1990s, space scientists experimented with a number of innovations that transformed the manner in which they managed complex undertakings. Through trial and error, they learned what worked and what did not.

The new management techniques require substantial in-house technical capability. Recent space science achievements have demonstrated that small teams made up of people intimately familiar with the spacecraft they design and fly can achieve performance goals at low cost if they work together within the same organization. Those teams require strong technical capability and they need to be assembled in such a manner so as to allow team members to communicate easily with one another.

NASA's secret weapon for completing Project Apollo arose from a combination of strong in-house technical capability and systems management techniques. The coordinating advantages afforded by systems management, along with the ability of technically competent NASA officials to "penetrate" contractors, allowed the agency to contract out most of the work of spacecraft and rocket assembly without losing project control. In the recent effort to cut the cost of human space flight endeavors, NASA officials have come to rely upon contractors while abandoning internal technical capability. This has created a "hollow" NASA, one with less capacity for either controlling costs or assuring reliability.

Many people agree that NASA has lost too much of its in-house technical capability, especially for human space flight. Agency employees who spend most of their time monitoring contracts cannot maintain the technical edge necessary to explore space. To produce outstanding results, they need to work with flight hardware. This necessity has been demonstrated repeatedly in the robotic flight programs, most recently within the Mars exploration effort. Successful endeavors, such as the Pathfinder and Mars Exploration Rovers, were led by persons with extensive "hands on" knowledge of spacecraft components. The spacecraft produced for those endeavors were assembled in-house under the direct supervision of team leaders. Experience suggests that 50 percent of the work (and money) associated with any successful, cost-constrained program needs to be retained in-house. Initial work on the Vision for Space Exploration began with a continuation of traditional contracting practices. Initial work on Project Apollo drew on old traditions as well, until cost-overruns and technical failures forced reform. The new management techniques that constrain cost while maintaining reliability require that project organizations be kept as simple as possible. When complexity rises, so do overall costs—often exponentially. Complexity arises from the desire for international cooperation and the tendency to spread work among many field centers. It arises from pressures to distribute project funds among a large number of contractors. Strictly speaking, these requirements are often irrelevant to mission objectives. The lean, uncomplicated management teams that have provided a recent alternative to those traditions are rooted in the "skunk works" philosophy advanced by Lockheed engineer Kelly Johnson. Widely adopted in modern industry, the approach depends upon the high level of interpersonal communication fostered on small project teams which acts as a substitute for the more expensive coordinating mechanisms contained in systems management.

Interestingly, those new management techniques do not require that individual projects be managed inside NASA. The Near Earth Asteroid Rendezvous (NEAR-Shoemaker) mission, for example, was managed using the same team-based techniques as guided the Pathfinder mission. Yet NEAR was carried out by a NASA contractor, the Applied Physics Laboratory, while Pathfinder was completed at NASA's Jet Propulsion Laboratory. Both are federally-funded research and development centers (FFRDCs), operated through parent universities. One of the most intriguing suggestions to emerge from the new initiative is the suggestion that all NASA centers be transformed into FFRDCs, thereby combining an allowance for contracting with the ability to expand in-house technical capability.

Both the Pathfinder and NEAR-Shoemaker missions were conducted by intact teams with very strong technical capability. In neither case was project responsibility split between separate groups, as doomed the performance of other low-cost missions like Mars Climate Orbiter. Divided project responsibility on cost-constrained missions conducted by small project teams substantially raises the probability of mission failure.

Summary

The challenge imposed by the Vision for Space Exploration could potentially transform America's civil space agency. It is hard to imagine humans traveling far beyond the Earth without such an institutional transformation. If the United States succeeds in returning humans to the Moon and sending humans to Mars, the agency that finishes the work will bear little resemblance to the organization that started it—just as the institution that landed humans on the Moon in 1969 hardly resembled the agency that received President John F. Kennedy's famous challenge in May, 1961.

Many of the practices that could be used to transform NASA already exist within the civil space effort, both in recent projects and past history. The methods used to carry out Project Apollo, the agency's original commitment to in-house technical capability, and the recent achievements on low-cost robotic programs all provide a strong foundation on which to construct the new transformation.

[1] In real year dollars, by 2002, U.S. taxpayers had spent $85 billion on the entire shuttle program. That is worth $137 billion in 2002 dollars or $29 billion in 1972 dollars. The shuttle program in its totality has cost 56 percent more than the initial $16.15 billion total cost estimate (1972 estimate) but flown only 19 percent of the original 580 flight manifest.

Dr. Howard McCurdy is professor of public affairs and chair of the public administration department at American University in Washington, D.C. An expert on space policy, he has produced six books on the U.S. space program, including the recently published *Faster, Better, Cheaper,* a critical analysis of cost-cutting initiatives in the U.S. space program. An earlier study of NASA's organizational culture, *Inside NASA*, won the 1994 Henry Adams prize for that year's best history on the federal government. He has also written *Space and the American Imagination* and co-edited *Spaceflight and the Myth of Presidential Leadership*. A native of the Pacific Northwest, Professor McCurdy received his bachelor's and master's degree from the University of Washington and his doctorate from Cornell University.

As you have seen, the question of how to develop a space economy is pivotal to developing a permanent human community on the Moon, and in fact, anywhere else we go in space. Nothing happens without money, and where that capital comes from is not nearly as important as how it is used –and how much of it goes into the development of infrastructure, both on the Moon and closer to Earth. Dr. Mike Ryan brings us several ideas and his focus is on how we develop this end of the Earth-Moon picture. The concepts of high risk waste disposal in space, and space solar power (to be beamed back to Earth) have been around since the early 1980's, where papers discussing them were often presented at Dr. Gerard K. O'Neill's Space Manufacturing conferences. What is interesting about Mike's essay is that it looks at activities and their cumulative effect on developing a space economy, and how it takes the discussion down to the very pragmatic risk versus profit consideration, based on a business person's sensibilities.

Making a Business Case for a Return to the Moon: Turning Dreams into Reality

by Dr. Mike H. Ryan

The Moon has always been a place of interest and excitement. If the explosive exuberance that science and technology had in the 1950s could have been translated into reality, people would have been living on the Moon by the mid-1970s. For many people who watched those first steps on the Moon in July 1969, it was hoped that they were the beginning of a new era of exploration and development, rather than a "planting the flag" exercise. While these expectations have not been fully realized, many believe returning to the Moon is a critical component of the human future.

The reality is that dreams are never free; there must be a way to pay for them. Some may suggest that when it comes to lunar operations, the sheer size and scope of proposed activities require government funds or government support and ultimately government control. After all, who but the government could afford to spend the vast sums of money required to establish and develop lunar operations? A presumption that the private sector could neither afford nor operate an ongoing lunar facility is often taken mistaken as the truth. What is frequently or conveniently ignored is that the governments, the United States' or any other's, record of success in establishing independent, self-governing, profitable, long-duration operations does not generate a great deal of confidence. The private sector has many advantages over its governmental counterparts, one of the most important is the ability to act without giving in to the less rational components of public policy driven by election cycle politics. Private enterprise has its own demons in the form of financial performance—generally measured by stock price—but there remains the opportunity to convince investors that a specific proposal has merit and the potential, if not the certainty, to become profitable. Lunar ventures must therefore stand on their own merits.

Lunar business opportunities must demonstrate that they can generate comparable revenues, profits, and investment returns commensurate with their terrestrial counterparts. Once a space-based venture shows the capacity to successfully generate sufficient revenue to cover its cost it will achieve a level of reality matching that of any other Earth-based operation. Therefore, the key for successful lunar ventures is quite simple; demonstrate the capability to generate revenue that will, in time, provide a comparable return to any that might be obtained elsewhere. If you do not expect a reasonable rate of return then as a private business it is unlikely that you would make the investment.

Today there are no scheduled passenger flights to the Moon, and neither UPS, Federal Express, nor DHL has the Moon on their routes. Barring some amazing new technology bordering on science fiction, the probability is quite low that businesses will be opening near the "Tranquility Base Memorial" anytime soon. Business people do not really want to hear about the assorted problems of launching material and people into orbit or anywhere else –they want to know when we can go, how much it will cost, and will it work the way they expect?

The transportation system that might make such things possible is fairly basic. First, you need a means to get people safely to and from Earth orbit on a regular schedule. You need the capacity to put very large things in orbit from time to time, preferably on demand. And, most importantly, you need to be able to put lots of smaller payloads into orbit virtually on a continuous basis. Moving supplies to the Moon requires that they be moved off-planet first.

Getting materials into LEO creates a serious problem for lunar development. Once they are there, they can be re-directed to the Moon for extended operations as required. Part of the difficulty in large-scale lunar business activity is that it needs a LEO market to justify the creation of the necessary transportation infrastructure. Re-supply of consumables such as water, food, spare parts, requires the reliability of a pipeline without the pipe. The regularity of maintaining an ongoing operation will require vast amounts of material with little, if any, interruption. You only need to think trains, planes, ships and trucks to realize that in order for business operations to proceed in a reasonable manner in space or vis-à-vis the Moon, transportation options are a serious part of the problem. It may be possible to get from here to there using a different set of options than our traditional space program may offer. The Russian space program has heavy lift capability—making it possible to put big things into orbit more or less on demand. Companies such as Scaled Composites, designer and builder of Spaceship One, are making some interesting inroads into the issue of getting people to and from Earth orbit less expensively and with reasonable frequency, starting with commercial sub-orbital flights. So two out of three characteristics needed to make business-rated transportation to low Earth orbit (LEO) a reasonable proposition are either available or under development. Building the requisite space pipeline for volume delivery of consumables and other supplies is more problematic.

Creating a space-based business system having the possibility of significant financial returns further improves the likelihood of developing the infrastructure to support other, more exciting, interesting or profitable projects.

Virtually everything that follows is either available in one form or another or could be built using known technology. Keep in mind is that this assessment is about creating a service with revenue potential that creates a specific need to be on the Moon. As it also happens, that service also provides for the development of a variety of orbital space infrastructures which in turn, provides a rational business need for operations to be carried out on the Moon, and the further transformation of lunar materials into goods needed elsewhere.

The amount of high-value hazardous waste available for disposal represents a source of revenue sufficient to justify the development of a space-based disposal system, and the hazardous nature of the waste makes permanent disposal desirable. The issue of public acceptance is arguably the most important driver in the development of any commercial space disposal venture. Any economic benefit of space disposal is moot if regulation and protest prevent operations. The effort that will be required in convincing people of the safety of this concept cannot be overstated. To provide the desired level of confidence with the acceptable level of safety in the event of failure, mass drivers (which are electromagnetic launch systems that hurl payloads without rockets) can be used for placing containers of material in a low Earth orbit. These containers would then be collected by manned orbiting collection stations. Containers would be loaded onto an appropriate transit vehicle that would be designed for a Sun intercept.

The number and placement of LEO stations would be determined by the launch sites and number of operating mass drivers. These stations, operating much as commercial diving or off-shore drilling rigs do now, would be manned by crews for a minimum 30-90 day shift. In fact, adapting hard suit systems currently used by diving companies for their commercial projects could make space-tending operations less expensive to develop. For example, there are at least two commercial grade hard suits presently available. Each is capable of prolonged periods of operation while keeping their operators at one atmosphere. The ability to keep development costs is a key component in moving this project or any similar project toward profitability.

LEO facilities need ongoing support and supplies; the same launch system used for waste disposal can easily provide containers of water, food, and other necessities. Re-supply of the orbiting collection points can be accomplished using containers sent up by the mass drivers. This should also reduce operating cost concerns and make continuous re-supply practical. More problematic would be crew changes. The current experience of the International Space Station illustrates what happens when transportation options become unexpectedly limited. If the development timetable for vehicles such as Spaceship One continue at the present pace it may be reasonable to see crew replacement vehicles sooner rather than later. Actual time in orbit is a question of access and relative cost. Longer periods between crew rotations are possible, but the tradeoffs include better quarters, more entertainment options and higher wages and/or salaries. As the system developed, it is probable that additional cost and revenue considerations would be pursued. Generally, the objective for most business operations is to reduce operating costs and whenever practical, augment revenues. One of the major costs of operation within the scenario is the electricity used by the mass drivers and their supporting industrial facilities. There are several approaches that might be considered depending on the

SPS Rectenna receiving power transmission from space.
Courtesy: Space Studies Institute, Princeton, New Jersey, USA.

location for generating additional electrical power to supplement or replace that purchased through standard commercial sources. Given that the probable preferred location for mass driver sites would be nearer the equator to take advantage of the Earth's rotation, opportunities to create power sources of sufficient size to offset their cost may be somewhat limited. Consequently, an additional opportunity in the form of building and using solar power satellites (hereafter referred to as SPS) can easily be envisioned as a follow-up to the waste disposal operation. SPS systems have the potential to be as cost effective as more traditional sources of electric power but with greater geographic flexibility.

Large-scale Orbital Space Construction
Courtesy: Space Studies Institute, Princeton, New Jersey, USA.

The practical arguments against SPS systems have often focused on the large amount of materials need for their construction, that must, by necessity, take place in orbit. Equally daunting is that many of the needed components would be quite large and difficult to launch from the Earth. And, even with the relative efficiency that might be obtained by the use of mass drivers to launch large numbers of small components the cost per SPS unit would be very high. The alternative use of lunar materials or lunar manufactured components could reduce the construction and manufacturing costs of SPS systems. The absence of an atmosphere when combined with the lower lunar gravity could create an ideal opportunity to manufacture SPS components on the Moon, and place them into the appropriate orbit using a Moon-based mass driver system. The cost saving that might be obtained through the use of lunar raw materials alone for the construction of large numbers of SPS systems should provide an economic justification for the creation and operation of a companion commercial lunar facility.

Lunar mass driver—moving lunar materials to LEO.
Courtesy: Space Studies Institute, Princeton, New Jersey, USA.

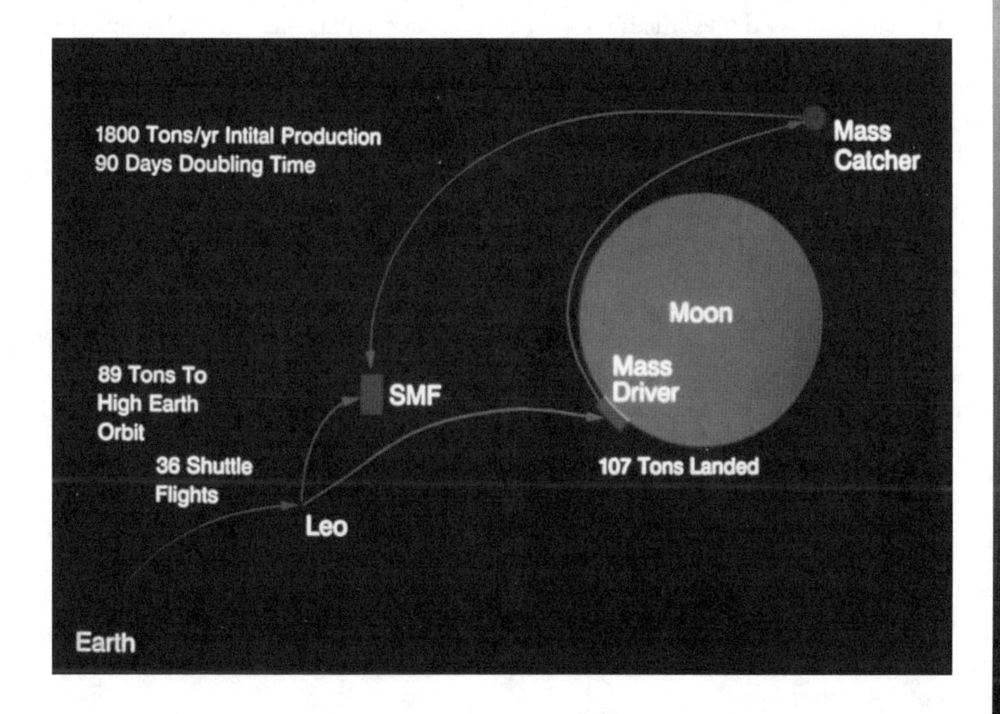

One possible pattern for lunar materials flow.
Courtesy: Space Studies Institute, Princeton, New Jersey, USA.

Backward integration is a common business practice in which manufacturers extend their operations back to the components or raw materials used in the construction of a specific product. Depending on the volume or size of a SPS construction operation, the availability of less costly lunar material could yield a commercial lunar presence, depending on the cost and the demand for electrical power. Although, the initial customers are government institutions, part of the appeal is that there are numerous nations with similar problems, thereby reducing the dependence on a single customer. In strategic terms, the number of waste suppliers diminishes the power that any single customer might attempt to exercise over the business.

A general analysis of hazardous waste disposal via a space-based system is contained in the Commercial Space Transportation Study (CSTS) produced in 1994. While many of the components of CSTS approach are viewed as problematic, its underlying arguments as to the need for disposal are essentially the same here. First, the longevity of the nuclear waste hazard represents a current and ongoing problem. Second, the terrestrial solutions for long-term storage of nuclear waste are still open to question. Third, the amount of money allocated to deal with the problem suggests significant revenue potential if a lower cost and/or more effective solution could be created. Lastly, there appear to be few competitors poised to enter the high level waste market apart from government institutions. Effectively, this translates into a market opportunity of considerable size, revenue positive, and of significant duration.

There is a tremendous stockpile of high-level nuclear waste in this country left over from 50 years of bomb building and 35 years of nuclear power generation. With the ending of the Cold War, there is additional plutonium to dispose of in the safest way possible. In addition, the nations which comprised the former Soviet Union have an abundance of high-level waste. In spite of issues relating to paying for disposal operations, there is no dispute about the need for high-level disposal options. Some estimates based on nuclear power generation data put the total worldwide high-level waste exceeding 100,000 metric tons. Some of that waste is currently being processed (glassified) for aboveground storage (e.g., in France), but most will be sitting in temporary storage tanks for the foreseeable future. In the US, the amount of nuclear waste to be disposed of is estimated to be in excess of 40,000 metric tons of spent nuclear fuel and another 10,000 metric tons of defense related wastes. The market for permanent disposal of high-level nuclear waste off-world is therefore of sufficient size to justify further interest.

Nuclear waste disposal is an expensive business. Much of the ongoing expense for the storage and treatment of nuclear waste within the U.S. is to be obtained from The Utility Waste Fund. The value of the fund is now in excess of $12 billion By the year 2030, nuclear power is expected to expand to about 190-250 gigawatts, generating more than $1 billion per year for the waste fund. Government estimates put the cost of the first permanent repository in the vicinity of $28 billion and the second in the $17 billion range ($45 billion total). These estimates probably represent the lower end of the cost envelope as other estimates go as high as $53 billion. However, using the estimated U.S. cost for repository disposal as a starting point estimate, each metric ton of waste (50,000 metric tons of U.S. waste only) translates to a cost structure of approximately $900,000 per metric ton. This does not include things such as ongoing security, oversight or any associated expenses. For instance, landlord activities for the sites in question could increase expenses an additional $13 billion for just the first third of the next century. Transportation to the sites is not included as transportation costs are subsumed in any waste disposal system. Regardless, the revenue potential from the safe and efficient handling of high-level waste should be sufficient to warrant investment in infrastructures, assuming an acceptable financial return.

Mass drivers have been discussed as launch systems for some time as the pioneering efforts of the Space Studies Institute attest. The technology is relatively straightforward. Design studies have gradually grown to include working examples constructed both by private institutions and the military. The fundamental issue is whether or not such a system can be scaled up to the size required to provide an efficient means of placing waste containers in a LEO. The technical limitations do not appear to be as daunting as site selection and public concern. Increasing the size of such a system to encompass the scale necessary for the described scenario would ideally result in a capability of placing, at least, metric ton containers in orbit.

Although waste disposal is a volume-based business, the sheer number of launches required to produce the desired revenue stream should not be viewed as excessive. At the same time, the point of using such a system is to provide a capability of launches twenty-four hours per day with many launches per hour. Without stretching credibility, a significant amount of material can be handled in a relatively short time period.

From a business perspective, placing all of your reliance on a single system would be considered imprudent. The more likely course of action would involve several mass drivers. The configuration is less important than the realization that redundancy for reducing business risk is not much different from redundancy to reduce technological risks. Regardless, additional launch capability increases the tonnage to LEO by whatever factor would make best economic sense. Scaled up mass driver costs can be compared to the creation of a new aircraft, only less costly. Development costs for new planes can easily approach $10-12 billion dollars or more. Optimistic projections for mass driver development have ranged from several hundred million dollars to more than two billion dollars. Investment thresholds at even the higher estimates could be met if waste disposal volume is increased or if the economies of scale provided by the system reduce the cost of each ton to orbit. Construction costs are more difficult to estimate because a significant part of the cost would be determined by the site selected for operations.

Presuming the desirability of a near-equatorial location with a relatively high terminal elevation, mass driver costs should approximate those of any other large scale projects in which most of the components can be prefabricated. Depending on its design, this system could have more in common with the Alaskan pipeline project than with more traditional space projects. For example, it may be possible and highly desirable to create mass driver sections using extruded concrete pipe technology. Even if located in very difficult terrain such a system could price out competitively with other launch alternatives but would have a significant advantage in both operational safety and in the frequency that payloads could be launched.

The point is to provide a permanent, yet economically-viable, approach for high-level waste disposal. Part of that approach is a desire for safe and reliable handling of the waste material. Such caution suggests that it is unwise to use rockets to place this material in LEO. In much the same vein, once in a LEO it seems reasonable to have individuals on the site to handle the waste and deal with any potential problems. Some argument can be made for telepresence in the handling of containers. Plans exist for the construction of orbiting habitats from external shuttle tanks, specially designed components shipped up from Earth, expandable modules, inflatable bubbles and many other approaches. The choice in this case is one of safety, utility and, ultimately, of cost. It need not be fancy, but the station (or stations) placed at collection points must be relatively self-contained and cost efficient. The best approach for station construction might be to get non traditional sources for its design. The operation of a LEO collection site capable of handling several hundred tons a day is not terribly complicated. The station is a basic cargo terminal. Each transit vehicle might hold several weeks' worth of canisters. When full, the vehicle is sent on its way. A crew for a station could consist of some 15-25 people per rotation. The overall effectiveness of the operation could be improved by increasing the scale as the amount of waste to be handled increased.

Assuming that a space-based waste disposal system becomes operational, it would be a relatively short time before efforts would be made to improve the rate of return and/or reduce operating costs. Investor expectations and/or concerns about operating efficiency would focus attention on improving the business within a short period. One clear avenue of inquiry would be to reduce the constant cost associated with electric power utilization for the mass drivers. A second would be to create peripheral business opportunities from the now available space infrastructure. Both areas could be addressed with the development and construction of solar power satellites. An SPS system creates the opportunity to provide electric power at a constant cost and reduces the dependence on outside suppliers. Conceivably, it could also out price the cost of electricity from other sources. These factors readily convert to competitive advantages in the face of competition from other disposal systems. It could also increase the entry barriers to competing launch systems.

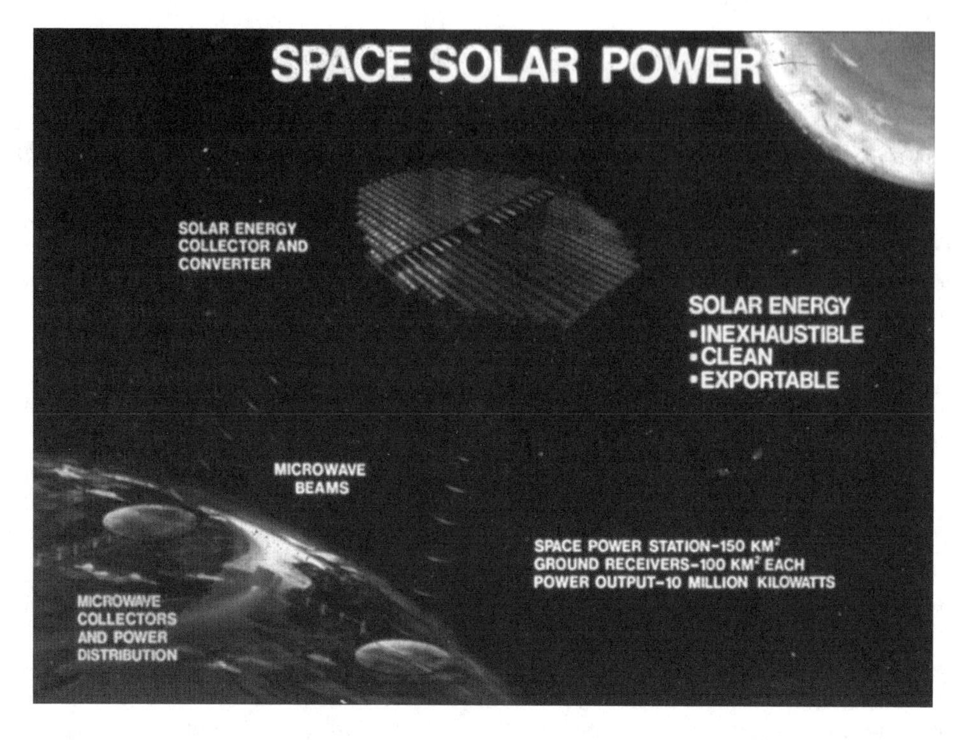

Summary of Space Solar Power Advantages.
Courtesy: Space Studies Institute, Princeton, New Jersey, USA.

In addition, the construction of an SPS power grid for internal use clearly creates the expertise to deploy such systems elsewhere. The demand for clean sources of electric power is likely to continue to grow throughout the twenty-first century. Electricity is and will continue to be the fastest growing component of energy demand. It is estimated that by 2020, total world electricity demand is expected to rise from 12 trillion kilowatt-hours to 23 trillion kilowatt-hours. However, it is among developing nations where SPS systems may have the greatest potential. Electricity demand is projected to grow at more than twice the rate of growth in the industrialized countries.

In the developing world, vast infusions of foreign capital will be required to sustain growth in electricity supply. Many developing countries are already coping with power shortages. The future power needs of such populous countries as Brazil, China, India, Indonesia, and Pakistan are immense, presenting investment demands beyond the financial means of their domestic capital markets or government resources. As a result, attracting foreign investment is critical to the successful expansion of the electric power sectors in these nations. Opportunities to use relatively less expensive, non-polluting sources, such as SPS systems, could be very attractive. In part, they could become financially attractive investments because of the availability of a supporting space-based infrastructure.

The focus of this scenario is that there are sources of revenue to develop and construct a space-based infrastructure. The businesses involved are not exactly exciting—garbage disposal and to a lesser degree production of electricity. These are basic utility operations common to all parts of the planet. The fact that they are common is part of their attractiveness for creating mainline business opportunities. Some parts of the approach are somewhat radical; however, the essential components represent mainstream technology and fall within current industrial capabilities. They are extensions of what we know combined with questions for which there are means of obtaining explicit answers. Is it possible to actually build the described infrastructure? The technical problems are of the "what will it cost variety." The political issues are mostly of the "if we can't do it here—there are other choices" contingent. Given the will, vision, investors, willing customers, and the essential know-how, it can be done. If nothing else, this scenario might help people to look for other opportunities upon which to build their space systems and lunar enterprises. As early ship-borne travelers noted, the ship took no notice of whether the voyage was one of high adventure to explore unknown lands, or a cargo of trade goods bound for sale on the home market. The point is that whether the ship was intended for exploration or commerce it provided the same service—transportation.

In the 30+ years since a human last walked on the Moon, many things have changed and yet surprisingly, some have not: our willingness as a species to confront the future with vigor and our willingness to embrace new and exciting technologies. A new generation of entrepreneurs continue building on the vision and dreams of those who dared to dream not only of people in space but of people actually living and working on the Moon. Regardless of whether those that dream learn to do or those that do learn to dream—*We will return to the Moon.*

1 A case could be made for a mass driver site that provides capability for polar orbits etc. making the scenario is a bit more complicated but this would not affect the rationale supporting orbital structures or lunar activities.

Dr. Mike H. Ryan is professor of management at Bellarmine University's W. Fielding Rubel School of Business, in Louisville, KY. Previously he founded and operated a multimedia company, Prometheus Press, which published and edited *Space Business Notes*, one of the Internet's first space business journals. Dr. Ryan earned MS and PhD degrees from the University of Texas at Dallas. He is the author of several books, more than 30 scholarly articles and serves as senior editor for an ongoing research book series, and has testified before the House Committee on Space, Science and Technology and the Texas Space Commission. In January 2005 he was elected to Fellowship in the British Interplanetary Society.

As I discussed in the preface, the big question facing this nation as we consider expanding outwards to the Moon, Mars and beyond is not technology, but psychology. We must change our way of thinking about space. More importantly we must change our way of thinking about who does what there. Many of the words in the following essay could have come from any one of a dozen leaders in the Frontier movement. Towards the end, when laying out his "Vision" Phil recites many of the same elements of the Frontier litany that we all share—and does it well. His tone is confrontational at times—the voice of one like many of us who have seen so many good ideas and people swallowed by an institution that has lost its way. We all hope that we are entering a new era, and that a new direction will be found by those entrusted with our tax dollars to help open space. We shall see if this is true. In the meantime, Dr. Chapman offers a bit of tough love to our space planners.

The Extraterrestrial Enterprise

by Philip K. Chapman, Sc.D.

When I was working on The Right Stuff, I realized that NASA had no philosophy of the exploration of space. We have never had a philosophy of why we are going into space, except to counter the Russians. At the beginning of the space program, that was enough... But when you reached 1969, 1970, and later, you could see what the lack of a philosophy has cost NASA.—Tom Wolfe, 1982

The Decline and Fall of the NASA Empire

When Jack Kennedy committed NASA to the Apollo program in May, 1961, U.S. experience in human spaceflight was limited to Al Shepard's suborbital hop three weeks earlier, which had lasted less than 16 minutes. The many unknowns included the physiological and psychological consequences of prolonged exposure to free fall, the feasibility of orbital rendezvous and docking, and the bearing strength of the Lunar surface. Kennedy's decision was an astonishing gamble, undertaken only to overcome the perceived Soviet lead in space, and to foreclose the possibility that the USSR would reach the Moon first and claim it as Soviet territory.

Despite the uncertainties, nearly everybody in NASA, from senior managers to the cleaning staff, believed that Apollo was nothing less than the first step toward a true interplanetary civilization. There was however very little discussion of a strategy for implementing that vision. The focus was on developing and testing the vehicles needed for the Lunar landing—i.e., on the means of spaceflight, not its objectives.

Grand aspirations were discouraged by Jim Webb, the NASA Administrator from 1961 to 1968, because he understood that the White House and the Congress were unwilling to spend more than was needed to meet the Soviet challenge. He believed that expressing "a philosophy of why we are going into space" would not yield more funding, but would supply ammunition to critics of the program, and he maintained that the proper purpose of planning at the agency level was not to propose future projects, but to define the steps needed to accomplish missions specified by the President. When Lyndon Johnson asked for his recommendations for the program after Apollo, he produced a list of options, without offering any criteria for choosing between them.

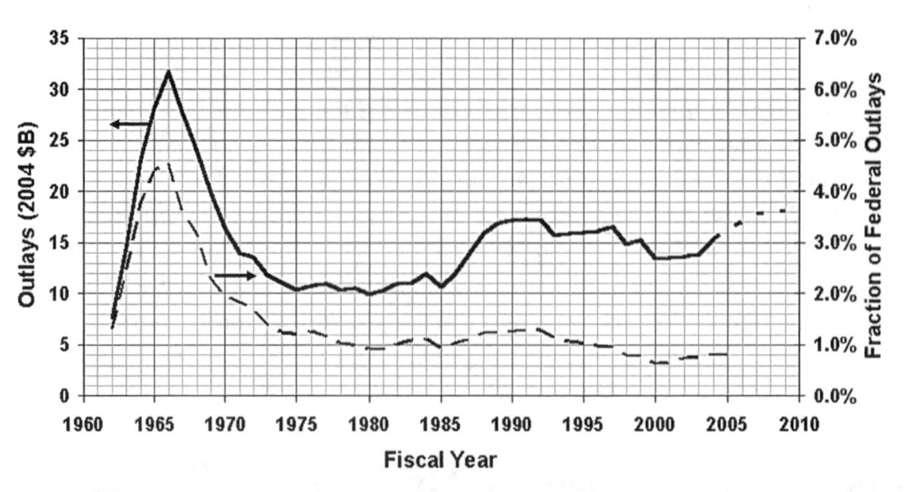

Figure 1: The NASA Budget History

In the mid-'Sixties, escalation of the war in VietNam and the introduction of domestic Great Society programs demanded cutbacks in other areas of the Federal budget. NASA was vulnerable, especially after the USSR accepted the Outer Space Treaty (OST) of 1967. This agreement banned military activities and national territorial claims on the Moon and thus undermined the rationale for Apollo. In 1968, the annual review of funding by the Bureau of the Budget noted that "the case for continuation of a manned space flight effort after Apollo is one of continuing to advance our capability to operate in space on a larger scale, for longer duration, for ultimate purposes that are unclear." As shown by the solid curve in Figure 1, NASA outlays peaked in 1966 at $31.8 billion[3] and then fell precipitously, to near $10 billion by 1975.

While challenges from other nations to US interests in space are inevitable, the particular tensions that led to Apollo are unlikely to recur. A need may develop for US military personnel in space, but there are few plausible scenarios under which a substantial civil program in human spaceflight may again seem vital to US national security.

Since NASA has failed to articulate any compelling alternative to Soviet competition as a reason for human spaceflight, the budget has never recovered. The *Challenger* accident in 1986 forced an increase, but funding drifted downward again under President Clinton. It leveled off in the current Administration, until an uptick after the *Columbia* accident. The proposed return to the Moon will require an increase, but there is no assurance that the Congress will provide the requested funds.

The dashed curve in the figure shows NASA outlays as a percentage of the Federal budget. By this measure, the priority of the space program among Federal concerns has decreased almost every year since the peak. The overall decline is by a factor of more than five.

Despite the cutbacks, NASA has spent more than $four75 billion (200four dollars) since Apollo 11. If spent wisely, as seed money to stimulate space development, that very large sum could have established a growing, self-sustaining extraterrestrial enterprise, offering opportunities for thousands of people to live and work off Earth. NASA chose instead to build the shuttle, which it claimed would provide reliable, cheap access to LEO.

NASA based shuttle cost projections on a wildly unrealistic economic model. First, the direct operating cost was projected to be less than $600/kg ($270/lb), a figure based on absurdly optimistic assumptions about maintenance requirements. Second, the mission model assumed that traffic to orbit in the 'Eighties would support 60 flights per year, with no justification except that this number was needed to permit an acceptable amortization of fixed costs.

Realizing that this traffic estimate was unconvincing, NASA compromised the design in an attempt to build a shuttle that could meet all launch needs, military, civilian and international, and then persuaded President Nixon to phase out production of expendable launch vehicles (ELVs). The result of this one-size-fits-all folly was a shuttle that is fragile, needlessly complex, unacceptably dangerous to fly and appallingly expensive to maintain.

Long before the maiden launch of the first shuttle (*Columbia)* in 1981, it was obvious that traffic to orbit would be an order of magnitude less than had been assumed. This was perhaps fortunate, since the turnaround time between flights proved to be three to six months, instead of the two weeks assumed in the mission model. Flying 60 times a year would have required a fleet of 20 to 30 shuttles, far beyond what NASA could afford. The plan had called for a ground crew of a few dozen people, but in fact an army of 10,000 technicians disassemble and rebuild each "reusable" shuttle after each flight. Given these realities, it was no longer possible to pretend that the shuttle could compete on cost with ELVs—but the United States had no alternative, since ELV production had ceased.

This blunder opened a window of opportunity for development of the French *Ariane* series of launch vehicles. As a result, the US market share of worldwide commercial launches fell from 100% in 1980 to 0% in 1986, and is still less than 25%. This loss is an indirect cost of the shuttle to the US economy.

Instead of learning the fundamental lesson from the shuttle debacle, which was that we cannot design effective space facilities if we have no strategic plan for their use, NASA compounded its error by committing itself to constructing another recklessly complex tool that had no credible purpose. The space station *Freedom* and its successor, the International Space Station (ISS) were claimed to be scientific facilities, but no research program was ever defined that could conceivably justify the annual cost, which exceeds the entire budget of the National Science Foundation (NSF).

In the original design, the ISS had a crew of seven, but cost overruns have forced deletion of a habitation module and a lifeboat that could return that crew to Earth in emergency. The shrunken station will accommodate only three astronauts, even when fully assembled. In normal operations, only one of them is American. The cutbacks also gutted the research program, by eliminating much of the scientific equipment aboard the station and reducing scheduled shuttle flights in support from six to four per year. This hardly matters, because the station is so complex that maintenance tasks occupy 90% of the time of three people, leaving very little opportunity for any kind of scientific work.

The ISS is an edifice whose unaffordable extravagance has consumed all available resources without creating any significant capability for useful work, so that it has no definable purpose except to serve as a destination for shuttle flights. We would not need the shuttle if we did not have the station, and we would not need the station if we did not need something for the shuttles to do. The entire human spaceflight program has thus become an exercise in futility.

Because of these egregious mistakes, we have less capability in human spaceflight now than in the early 'Seventies. We could land on the Moon then, and we had a space station called *Skylab*, with berths for three astronauts (until NASA let it reenter and break up over Western Australia). A second *Skylab* was built and could have become the Earth terminal of a Lunar transportation system, but it was never flown. It is now a tourist attraction at the Air and Space

Museum in Washington, and the Saturn V to launch it is nothing more than a monstrous lawn ornament, moldering on its side at Johnson Space Center (JSC).

After wasting three decades (and a perfectly good Cold War), squandering hundreds of billions of dollars, frustrating the dreams of a whole generation of space enthusiasts and killing 1four astronauts, NASA's net achievement in human spaceflight is another three-person space station, 20 times more expensive than *Skylab* but with much less utility for research.

NASA was once a splendid example of what makes this nation great, but the loss of the sense of mission that energized Apollo has ruined the agency. We, the American people, gave NASA the historic privilege and momentous responsibility of creating a spacefaring society. It has failed the challenge.

The ISS and the Bush Initiative

In January, 2004, President Bush directed NASA to undertake a return to the Moon, with the objective of establishing permanent bases there, and to begin planning a manned mission to Mars. This initiative is a major improvement over running around in low-orbit circles, going nowhere. It offers NASA a possibility of redemption, but it cannot succeed without fundamental changes in approach.

The first problem is that the shuttle and the ISS are irrelevant to the Lunar mission, but they remain such a major drain on resources that they cripple NASA's ability to pursue the Presidential directive. NASA now plans to retire the shuttles in 2010, after about 20 more missions to finish assembly of the three-person "core complete" version of the ISS. The station is scheduled to remain in operation until 2016, supported after 2010 by unspecified "alternate transport."

Based on experience to date (two losses in 113 missions), the probability of losing at least one more shuttle during the remaining missions is 30%. In other words, completing the ISS is like a game of Russian roulette using a revolver with two out of six chambers loaded. Only the most critical national needs could justify taking such a risk with the lives of astronauts, and the ISS certainly does not qualify.

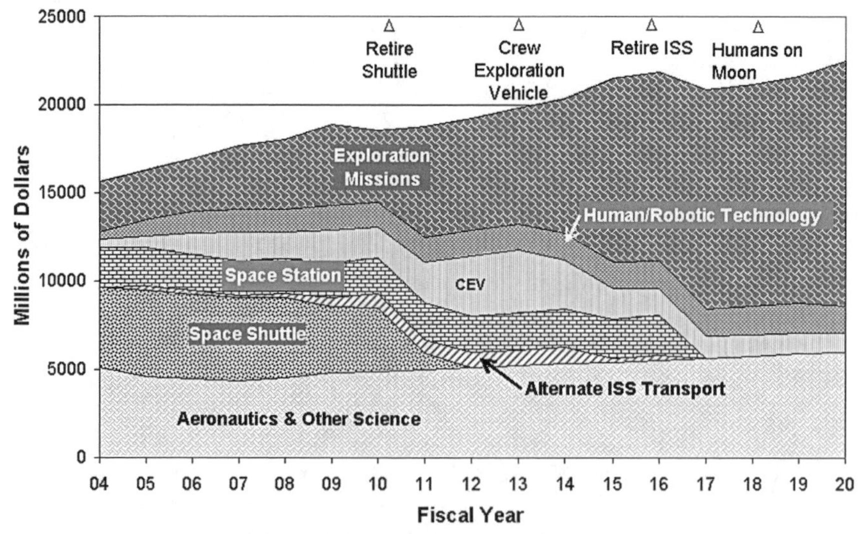

Figure 2: NASA Budget Proposal

Figure 2 shows the NASA budget proposal for the Bush Initiative. The total indicated expenditure on the shuttle from 2005 through 2010 is $26.1 billion. Since 20 flights are scheduled, the average cost of each one will be $1.3 billion. The maximum payload of the shuttle to the ISS is 17,100 kg (37,670 lb), so the minimum average launch cost to the station will be $76,300/kg ($3four,600/lb). This is more than five times the current price of gold, a truly incredible charge for freight to a useless facility.

According to the figure, the expenditure for the ISS and its support will exceed that for the Lunar program through 2010. The cumulative expenditures from 2005 through decommissioning of the ISS amount to $27.1 billion for the shuttle (including termination costs), $6.2 billion for "alternate transport", and $25.2 billion for the station itself, for a total of $58.5 billion. What will we gain from this very large investment? NASA has tried to give the ISS some semblance of relevance by claiming that its purpose is now research concerning human factors in long-duration spaceflight (such as a Mars mission). The obvious advantage is that using the crew themselves as guinea pigs allows data collection even though they are spending their time on maintenance rather than productive scientific work, but the new policy has serious problems:

> The proposal may violate NASA's contractual obligations to the international partners in the station, by eliminating the last vestiges of the research that was the reason for their investment.
>
> The new priorities amount to a tacit admission that completing and maintaining the station is a monstrously expensive make-work project that merely gives the crew something to do while the medics observe their reactions to the environment. Why not have the astronauts work on something useful, such as the return to the Moon?
>
> There is in any case no conceivable justification for these studies aboard the ISS, since the Lunar project itself will offer much better opportunities for the research, at very little incremental cost.

We already have experience with 18 people (fourteen Russians, three Americans, and one German) who have spent six months or more in free fall. Valeri Polyakov holds the record, with a total of 680 days on Mir, or almost two years (242 days in 1989 and another 438 during a second visit, starting in 1995). Does planning the Mars mission really require more data on the physiological effects of microgravity? If it does, routine Lunar operations will almost certainly involve a modest space station as a transfer node in the transportation system, which could also serve as a laboratory for such studies. The most likely location is L1[5], the nearside Lunar Lagrangian point, where passengers and freight from Earth would transfer to a Lunar landing shuttle.

In any case, the most significant hazard en route to Mars is not free fall, but cosmic and solar radiation in interplanetary space. The ISS is of limited utility for studying this issue, because it is protected by the magnetic field of the Earth, but astronauts on or near the Moon will be exposed to the full spectrum. In particular, a facility at L1 would be ideal for radiation studies, and for development of shielding techniques. The other major area for study concerns the psychological and sociological effects of confinement and isolation. On the ISS, the Earth is always just outside the window, but the Moon, like Mars, is another world, where we can learn how people cope with extreme physical and psychological separation from home.

Since the ISS serves no discernible purpose, the rational course is to cut our losses. Outright cancellation may be too politically embarrassing to contemplate, and might require extensive negotiations concerning compensation of the international partners for their lost investment in the project. On the other hand, wasting another $58.5 billion of the taxpayers' money is not a viable option, especially when that money is sorely needed by the Lunar program.

The *Columbia* accident is persuasive evidence that the shuttles are too old and too dangerous to keep flying. Completing the ISS entails unacceptable risks, and that fact offers a face-saving escape from the dilemma. The only prudent course is to mothball the station, move it to a higher, stable orbit, ground the shuttles permanently after the minimum number of flights needed to accomplish these tasks, and defer indefinitely decisions about reactivation or cannibalization of the structure. This would save at least $50 billion, greatly improving the feasibility of the return to the Moon.

Justifying Spaceflight

The immense public enthusiasm aroused by Apollo 11 was not due to US technological prowess, nor to the victory in the race with the Soviet Union. It was certainly not fascination with Lunar mineralogy (in fact, the geological emphasis in later missions baffled and bored most people). The appeal was of course due to the dramatic import of our first visit to another world. Armstrong and Aldrin seemed, like Lewis and Clark, the vanguard of an historic expansion.

Everybody understands the true significance of human spaceflight: it represents our first steps outward into a realm far larger than this small, fragile planet. We are fortunate to live during the first stages of the process that will bring life to barren worlds in the solar system, and eventually across the galaxy. Many NASA managers, perhaps most, share this vision, but actual steps toward implementing it lie far beyond their planning horizon.

The principal reason for this myopia is that no foreseeable NASA budget is even remotely adequate for the task. In fact, *any* human spaceflight program controlled by NASA is doomed to insignificance, because funding is determined not by the results achieved, but by politics, in competition with other federal priorities. Sustained growth is unthinkable, no matter how successful the program may be, so only a tiny group of astronauts can ever make it into space.

The general expectation among NASA people is that the galactic diaspora will happen some day, but not on their watch. In the interim, talk of an interplanetary civilization is discouraged, because it sounds like science fiction and raises expectations NASA cannot fulfill. The best NASA can offer are occasional Giant Leaps for Mankind—i.e., spectacular but essentially symbolic feats, such as the first human landing on the Moon (or Mars)—in the unreasonable hope that such stunts will persuade American taxpayers to continue supplying billions of dollars. Some engineers, even more delusional, seem to think that sheer technological complexity, as in the shuttle and ISS, is sufficient by itself to generate public enthusiasm.

The official NASA position is that the objectives of human spaceflight are to explore new worlds and to pursue scientific research. The implicit conception is that the solar system is an arena something like Antarctica, where the advance of knowledge is adequately served by small, temporary expeditions, but this analogy between spaceflight and terrestrial exploration breaks down when examined. The acquisition of knowledge about the solar system is a laudable goal, but in human spaceflight it cannot be more than an ancillary function in a program that is justified for other reasons.

One major difference is the sheer size of the solar system. The surface area of the Moon is a quarter of the total land area on Earth, or three times that of the Antarctic continent. There are four other planets with solid surfaces (i.e., excluding the gas giants), plus 19 other moons and four asteroids that are large enough (>400 km in diameter) so that gravity has made them spherical. The total surface area of all these bodies is eight times the land area on Earth. Human exploration of all this territory will take millennia, at the maximum pace that NASA can manage.

The second issue is the cost of life support. It is generally cheaper to take people to Antarctica and to support them there than to build robotic systems that could operate in that environment, but that is not true in space (under present conditions). There is no chance at all that NASA can convince the scientific community that the research productivity of astronauts justifies the cost of supporting them, when there is still so much that we can learn, at much lower cost, using unmanned systems.

Finally, if human spaceflight is purely a scientific enterprise, the competition for funding must eventually be with other uses for scarce research dollars, and not with Federal priorities in general. The NASA budget is less than half the peak in the glory days of Apollo, but it still amounts to $15.four billion. While this expenditure covers aeronautics as well as space studies, it is far above the level expected for scientific programs. The NSF spends only $4.1 billion on research in *all* the sciences.

At the very least, we should expect a bushel of Nobel prizes from the investment in human spaceflight, but the research results from the shuttle and ISS have been pathetic (especially when compared to the achievements of unmanned observatories and planetary probes during the same period). The experiments have been perfunctory, of a kind that might win prizes at a high school science fair, and they have not prompted a single patentable idea by any astronaut.

Table 1 The National Science Foundation Research Budget, FY 200four ($Millions)

Biological Sciences	562.22
Information Technology	584.26
Engineering	536.57
Geosciences	687.92
Math & Physical Sciences	1,061.27
Social,Behavioral & Economic Sciences	211.74
Polar Research	329.93
Integrative Activities	132.45
Total	$4,106.36

For comparison, Table 1 shows the FY 2004 research budgets for the areas supported by the NSF. In particular, the annual budget of the US Antarctic Program is $300 million, including logistic support (the rest of the Polar Research line item covers Arctic studies). This modest funding is sufficient to support 270 people on the Ice in winter and 1250 in summer. NASA spends 50 times more money, but the crews it can deploy in space are two orders of magnitude smaller than the US teams in Antarctica.

The glamour of the proposed return to the Moon, coupled with appeals to the political pork barrel, may lead to small increases in funding during the next few years, but the prognosis must be that NASA's reliance on science as justification for human spaceflight amounts to slow institutional suicide. American taxpayers and their representatives in Congress have shown remarkable patience, but it is far too much to expect that they will tolerate the present paltry progress for decades to come. Without a radical change in NASA's approach, opponents will whittle away at the budget until it reaches equilibrium with other research, at a level well below funding for the NSF. Human spaceflight will of course cease long before then.

The Bush Initiative has brought the human spaceflight program to a crossroads. If we continue down the path that NASA has been following, we can expect more of the waste, mismanagement, exorbitant costs and feeble performance that have characterized the program for the last three decades. The proposed Lunar base will be, at best, another symbolic gesture, expressing the forlorn hope that we will establish genuine Lunar settlements at some time in the unspecified distant future, before the sun dies. The project will offer no real progress toward a spacefaring society, but will instead waste another 20 years before we can reform the system. If a Martian landing follows the Lunar base and is managed according to present NASA practice, the delay will be 50 years.

NASA will not last that long. The prospect down this road is bleak indeed, for spaceflight and for NASA.

The obvious alternative is to abandon the self-inflicted limitations of the NASA program. Now is the time to change direction, before vested interests have developed in the standard NASA approach to the Bush Initiative and while the space budget can still supply seed money.

If human spaceflight is to continue, NASA must accept expansion of our society into space, not merely exploration, as the primary long-term objective. We need to begin an extraterrestrial enterprise that will allow an escape from the environmental constraints of the Earth, remove barriers to economic progress, provide room for diverse cultures to flourish without mutual antagonism, and offer young people of all nations opportunities for adventure and achievement. Commitment to these goals and demonstrable progress toward them are essential if NASA is to regain support from the public and from politicians.

Converting the solar system into an abode of life is a colossal task, but not more daunting than that facing the first Europeans in the Americas—where, five hundred years after Columbus set foot in the New World, the United States has become the strongest and richest nation on Earth. Technology gives us immense advantages compared to the early American colonists, but it will take centuries to tame whole worlds. The process does not require inordinate public expenditures, because the only essential requirement for exponential growth is creation of a positive feedback mechanism. In other words, we need to establish an organizational structure such that space operations can themselves generate the revenues needed for expansion.

There can be cities on the Moon and Mars, and among the asteroids, long before the fifth centenary of Apollo 11, but nothing will happen if we don't get started.

The Road to Other Worlds

We already have a societal institution that is expressly designed to enable growth through positive feedback (and also to limit the potential damage in risky operations). The profit-making, limited-liability joint stock corporation was invented during the 17th Century for the specific purpose of raising the capital needed for expeditionary ventures. The first British settlements in North America, at Jamestown and Plymouth, were organized by the Virginia Companies of London and Plymouth, under charter from King James I. This is a tool ideally suited to the needs of spaceflight: why are we not using it?

The key to our future in space is to shift principal responsibility for human spaceflight from NASA to the private sector.

> It should be obvious that investors, not taxpayers, must provide most of the funding for a healthy, growing space program. There are many reasons:
>
> Growth in the NASA budget is an unacceptable drain on Federal revenues, but growth in the private sector is a boon to the economy.
>
> Private participation in space operations can stimulate exponential growth because

it permits reinvestment of profits in expansion, a mechanism unavailable to NASA.

Corporations can make much more rapid progress than NASA, at much lower cost, because they can take risks that government agencies cannot.

Commercialization can convert the space enterprise into a source of tax revenues instead of a sink for them. After more than four decades of public support, isn't it time that human spaceflight started paying its own way, instead of contributing to the Federal deficit?

Private ventures are not nearly as susceptible as NASA to pork barrel politics and pressures.

A growing commercial program would create the constituency needed to maintain Federal funding.

The purpose of human spaceflight is to open space to humans—i.e., to all of us, not just government employees. The appeal of the program depends on the perception that it is opening a new frontier where people can escape the increasing constraints of life on Earth. A program controlled exclusively by government is incompatible with that perception. It cannot survive because it contradicts a principal reason for popular support.

International cooperation between national space programs is naturally limited to space-capable nations, excluding most of the Third World countries that control the UN General Assembly. Corporations need have no concern about the nationality of their employees, so a commercial program can offer spaceflight opportunities to people of all nations.

Human spaceflight can be a potent demonstration of U.S. leadership, but a centrally planned, government-run operation sends the wrong message to nations struggling with the transition from command economies to democracy and free enterprise.

Engaging the engines of free enterprise does not mean eliminating government participation. The proper functions of government are to provide the initial funding that will get the process started; to overcome ideological, bureaucratic and technical obstacles to private operations in space; and to make it possible for corporations to profit from human spaceflight. Specific areas include:

Support for development of enabling technology;

Sponsorship of facilities such as simulators and test chambers, available for rent by anybody;

Funding for exploration and scientific research;

Utilization of Federal buying power in creating initial markets for products and services;

Incentives to overcome barriers to investment;

Development of a legal framework for acquiring, regulating and protecting private property and other rights in space;

Negotiations leading to international agreements that benefit US industry; law enforcement;

Search and rescue;

Traffic and debris control;

Protection of fragile environments in space;

Military applications of space technology; and

Provision for the security and, if necessary, the physical defense of US space assets and interests, public and private.

The Phases of Extraterrestrial Growth Nobody can forecast in detail how an interplanetary society will evolve over the next several centuries, but the general outline is clear and the initial steps are well understood. The probable development of an extraterrestrial economy includes four overlapping phases.

In Phase I, space systems owned and operated by government are built by private corporations to government specifications. This regime is useful in developing the technical capabilities of industry, but it offers no mechanism for growth and provides few incentives for cost control. In this phase, companies are risk-averse contractors, not adventurous entrepreneurs. Since external political factors determine the budget, contractors are engaged in a zero-sum game. Profit margins are regulated, so individual companies can increase their income only by maximizing the size of their contracts, at the expense of other contractors. Politicians encourage this process, because they are more interested in creating jobs in their districts than they are in curbing waste. The inevitable result, as shown by the NASA program, is that costs rise to levels limited only by public outrage.

In Phase II, which should have begun 30 years ago, directly after Skylab, government uses its purchasing power to attract investment from private sources and to create a market for space systems that corporations design, build, own and operate. It is not sufficient merely to replace NASA employees with contractor employees, supervised by NASA. The objective is to create substantial, independent, profitable corporations with experience in space operations, so the involvement of government must be limited to establishing broad policy guidelines. Incentives for investment may include tax breaks; grants to support technology development; market guarantees, anchor tenancy by government agencies in private space facilities; take-or-pay contracts for goods and services; prizes for achieving specified capabilities; public assumption of some downside risk; and grants of rights to extraterrestrial resources.

In Phase III, corporations begin to produce goods and services in space for sale in terrestrial markets. The resulting revenues directly expand funding for spaceflight and stimulate further investment. Analyses of commercial space operations often stop at this phase, apparently assuming that the only purpose is to serve earthbound consumers. This is like claiming that the export of pineapples is the only activity that makes economic sense in Hawaii.

The take-off point will be reached in Phase IV, when space becomes an independent economic realm—i.e., when trade between corporations and other entities working in space reaches a level comparable to trade with Earth. Imports from and exports to Earth will remain essential even when extraterrestrial settlements are well established, but economic activity off planet will surely grow until space-based products find their principal markets in space, not on Earth.

The Bush Initiative is an opportunity to put aside the mistakes of the past, and to begin at last the transition from Phase I to Phase II. A serious commitment to engaging the private sector can reduce the cost of the program to the taxpayers, begin development of the orbital infrastructure needed by the extraterrestrial enterprise, and open some initial commercial markets for privately funded human spaceflight.

Reforming NASA

NASA has tried on occasion to help private industry develop space capabilities, but the results have generally been disastrous. Numerous attempts to "reinvent NASA" have failed to yield any noticeable improvement in management or policies. The stubborn refusal to learn from catastrophic mistakes and the repeated failure of reform efforts suggest strongly that NASA is incorrigible. The sad truth is that many of the space enthusiasts who should be NASA's primary constituency have come to believe that the agency is an impediment rather than an aid to space development, and that the only solution is to shut it down and start over with a wholly new organization.

When President Bush proposed the return to the Moon, he also convened a prestigious Commission, under the chairmanship of former USAF Secretary Edward C. "Pete" Aldridge, Jr, to specify the steps needed to fulfill the vision. The Aldridge Commission Report, published in June, 2004, proposes sweeping changes in NASA. In particular, the Commission recommended that "NASA recognize and implement a far larger presence of private industry in space operations with the specific goal of allowing private industry to assume the primary role of providing services to NASA, and most immediately in accessing low-Earth orbit. In NASA decisions, the preferred choice for operational activities must be competitively awarded contracts with private and non-profit organizations and NASA's role must be limited to only those areas where there is irrefutable demonstration that only government can perform the proposed activity." Another recommendation was that the Congress "increase the potential for commercial opportunities related to the national space exploration vision by providing incentives for entrepreneurial investment in space, by creating significant monetary prizes for the accomplishment of space missions and/or technology developments and by assuring appropriate property rights for those who seek to develop space resources and infrastructure."

It remains to be seen whether NASA can accept this vision and pursue it without reservation. Some directorates within NASA are certainly trying. In particular, the new Office of Exploration Systems (OExS) in NASA HQ has sought the advice of the private sector in developing plans for the return to the Moon.

The implications of this new paradigm are however deeply disturbing to many old NASA hands, who will resist it by every means available, from bureaucratic inertia to outright rebellion. While NASA may establish the objectives of the Lunar program, the new approach means that most decisions about implementation should be left to entrepreneurs, giving the agency much less control over space operations. Moreover, many functions now served by NASA astronauts will be taken over by corporate employees, who will not be subject to the rigid control exercised by the Astronaut Office.

In the long run, growth in private operations in space will make NASA a minor, but still important player, in the enterprise. NASA may take the lead in occasional missions aimed at exploration of new territory, but its primary role will be like that played in aviation by its predecessor, the National Advisory Committee on Aviation (NACA). That very successful research organization provided much of the knowledge base that brought us from the Wright Flyer to the Boeing 747 in 65 years, but it did not try to run airlines. NASA should look forward to supplying analogous services for human spaceflight, as well as proposing and funding measures to jump-start the extraterrestrial enterprise.

Transforming the institutional culture in NASA is a difficult task, but failure in this endeavor means continued attrition of the budget. If NASA can adopt a role supporting private operations in space, and can pursue it with dedication and enthusiasm, then we can look forward to an historic expansion outward from our home planet, as *homo sapiens* becomes a spacefaring species. If the agency cannot, the Congress will have no option but to set up a new organization for that purpose. It would be a pity to lose the many dedicated people still working for NASA, or to disrupt its technical facilities, but the human future off Earth is much more important than preservation of the current institutional structure.

1 Biography (see below)
2 Budget Bureau, October 30, 1968. Reprinted in Exploting the Unknown: Selected Documents in the History of the U.S. Civil Space Program, NASA SP-4407, Volume I, pp. 495-499.
3 All figures are adjusted for inflation to 2004 dollars.
4 Data from the NASA website, at www.nasa.gov/pdf/54873main_budget_chart_14jan04.pdf
5 L1 is the point where the gravitational pull of the Earth just balances that of the Moon and the centrifugal force due to rotation of the Earth-Moon system. It is approximately 56,000 km above the nearside lunar surface.
6 In a speech in February, 2000, Marguerite Broadwell, ISS Commercial Development Manager at NASA HQ, pointed with pride to the fact that, in the history of the shuttle, "no astronaut has become an inventor as a result of performing these experiments." She thought this would reassure commercial interests who might sponsor experiments aboard the ISS but were worried about protection of their intellectual property. She also recommended "fully scripting the experiment" (i.e., giving the astronaut a checklist to follow that obviates any thought about what he or she is doing), or using "a self-contained unit" (i.e., an automated experiment in which the astronaut has no function). Her full text is online at http://commercial.hq.nasa.gov/files/staifpapers/intellectual_property.pdf . By any reasonable standard, the failure to obtain any value from human capabilities is a shameful disgrace, not an achievement.
7 Source: http://www.nsf.gov/bfa/bud/fy2004/tables.htm#technical
8 The total population of Antarctica, including all nationalities, is approximately 1000 in winter and 4000 in the summer. The median population at each base is 20 in winter and 45 in summer.
9 An example: In the late 'Nineties, several small companies, financed by investors, demonstrated substantial progress in developing cheap launch vehicles. NASA decided to help by funding a comparable but much more expensive project at Lockheed Martin, called the X-33. Since investors were unwilling to compete with NASA, funding for the small companies evaporated overnight. In 2001, after wasting $912 million, NASA canceled the project. By that time, some of the small ventures were bankrupt. However well-intentioned it may have been, the net effect of the X-33 was to crush private enterprise.
10 In "A Journey to Inspire, Innovate and Discover," Report of the President's Commission on Implementation of United States Space Exploration Policy, at http://govinfo.library.unt.edu/moontomars/docs/M2MReportScreenFinal.pdf
11 t/Space is one of several small companies currently under contract to OExS, studying architectures for lunar development in which the private sector is responsible for logistic support, including operation of the cislunar transportation system.

Phil Chapman was born in Melbourne, Australia. He wintered in Antarctica during the IGY, mostly at a remote, two-man camp, studying the aurora australis. After earning a PhD in physics at MIT, he was selected by NASA as one of the second intake of scientist astronauts, and served as Mission Scientist for Apollo 14. He has been active in the space advocacy community throughout his long career, including serving as president of the L5 Society (now the National Space Society) in the early 1980s. His principal research interests have included laser propulsion, solar power satellites and economical launch vehicles.

As you have seen, this book is slanted much more in the direction of the philosophical "Why?"and "How?" of our return to the Moon than the technological solutions or neat spacecraft and hardware that will be involved (although I have thrown in a light smattering of such discussions and some fun pictures.) It has been crafted this way because these sorts of decisions are what will drive the technology. You will also have noticed by now that some themes seem to re-appear again and again. Property rights, private sector leadership, support for individual initiatives and a benevolent if not catalytic role for government as time goes on and we move from exploration to settlement are all keys to our potential success or failure establishing ourselves beyond the Earth. Not only does each one of these "cultural" or strategic decisions right now effect the technologies we will use in space, they set into motion patterns of future behaviour and actions which can grow over time into dramatic differences in the societies we are seeding today. As an earlier writer pointed out, although the people in NASA, business and politics may each think they are acting to further their own goals, and that those actions are occurring in a microcosm, they are creating vectors which will ripple through time. Therefore decisions about who owns what on the Moon, the relationship between NASA and private sector firms in the transportation and service arenas are far more important than just "deals" and paychecks. They are creating a real roadmap for our future. Robert Zimmerman lays out too stark and clear examples of how that "future effect" can play out. Take heed.

Brave New World?

American Colonial History as a Guide for Designing the New American Space Initiative

by Robert Zimmerman

After two years in the New World, the Pilgrim colony at Plymouth still hovered on the edge of starvation. Supplies from England were scarce, and though the 100 or so settlers had learned how to grow Indian corn, they were still unable to produce enough food to feed themselves. Making the situation worse was the arrival of more settlers, most of whom came without provisions.

Then, the 1622 harvest was a failure, and an attempt to borrow corn from the Indians was rebuffed. To survive the winter Governor William Bradford was forced to place everyone on half rations. In desperation, some Pilgrims even became servants to the Indians, cutting wood and fetching water, while "others fell to plain stealing, both night and day."

"It may be thought strange that these people should fall to these extremities in so short a time," wrote Bradford. The Pilgrim settlers came mostly from the English middle-class, were literate and well educated, and were skilled workers. Yet, after three years they were barely able to provide for themselves, and the colony seemed on the brink of failure.

But the Pilgrim colony did not fail. Arriving on a continent with no homes, farms, roads, markets, or any of the conveniences these European settlers were used to, the Pilgrims were trying to accomplish the same task advocated by many space activists: build a new society in an empty and uncivilized wilderness such as the Moon and Mars.

What did the Pilgrims do to turn their effort around? In fact, what made each of the English colonies in America work, and what made some work better than others?

On January 14, 2004, President George Bush announced a new and bold space initiative. After completing the International Space Station and retiring the space shuttle, NASA is to embark on an ambitious long term effort to explore the solar system. First American astronauts will build extended and long term colonies on the Moon, and then use the knowledge gained from those colonies to extend human reach outward to Mars and beyond.

"Establishing an extended human presence on the Moon could vastly reduce the costs of further space exploration, making possible ever more ambitious missions," the President remarked that day. "We can use our time on the Moon to develop and test new approaches and technologies and systems that will allow us to function in other, more challenging environments."

To make these dreams happen, however, it would be wise for Americans to look back at our own history. The subtle differences in how colonies are established can cascade down through the centuries in ways that no one ever expects. Consider Virginia and New England. Both were settled by British citizens at the same time. Yet, one became a slave state, while the other was where the abolitionist movement was strongest. These differences led directly to the Civil War, and still affect the American scene today.

It is therefore imperative for elected representatives and NASA officials to understand how their decisions today will shape the future, for good or ill, both here on Earth as well as on the Moon and Mars, as they give birth to the President's space initiative.

As planned, will these first lunar colonies make possible a space society that is just and humane, or will they instead lead to evils like slavery?

Moreover, can we use the lessons of the past to encourage the establishment of prosperous and civilized space colonies, *as quickly as possible?*

Property rights

In Plymouth, the solution was amazingly simple. The Pilgrim colony had originally been structured as a joint stock company, whereby investors provided the funds to transport and supply the settlers, who in turn agreed to work for the company. For the first seven years of their settlement in America, all profits, property, and even their labor belonged to the corporation, of which the settlers were shareholders. Hence, all food, housing, and clothing was distributed equally among the settlers, regardless of skill, rank, or enterprise, making Plymouth not unlike a communist state.

Not surprisingly this arrangement did not work. Not only were the Pilgrims starving, but they failed to provide the investors in England with any profit. Men resented how those less capable or productive earned the same, while women resented doing housework for other men. Bradford found that neither encouragement nor force could make them work hard enough to produce anything for either themselves or the company. "They deemed it a kind of slavery," the governor noted, and thus found ways to avoid work.

After two years of bad crops and insufficient food, the colony's leaders decided it was time for a change. Each family was thus assigned a parcel of land for their own profit and use.

"This had very good success," wrote Bradford, "for it made all hands very industrious, so as much more corn was planted then otherwise."

Now even women who previously said they were too weak to work in the fields eagerly went to work, and were even willing to get their children to come with them to help. The colony soon had plenty of food, and except for periods of drought, never experienced a serious shortage again. In fact, the colonists were able to produce large surpluses, which they used for trade.

Life in the colony soon resembled life in England, which encouraged other Englishmen to consider immigration. In 1630 the great migration of Puritans began, bringing over 30,000 settlers to the Massachusetts area in less than a decade.

The rest, as they say, was history.

More Than Property

We would be grossly mistaken, however, to assume that property rights alone will guarantee the prosperity of any future space colony. While it might bring profits, success is not measured simply by the bottom line. Consider Virginia. Though it was more populous than New England, it was a miserable and wretched place to live, and became the home of slavery, what Alexis de Tocqueville called "the accursed germ" on the soil of America.

Just like in New England, the Virginia colony was begun as a joint stock company whereby all profits went to the company rather than the settlers. And just like New England, Virginia did badly under these rules. During its first five years of existence the colony experienced starvation, violence, squabbling, and bankruptcy. During the winter of 1609-10 — called "the starving time" — the population shrank from 500 to 60 due to famine.

Then, just like Plymouth, the settlers were given their own land to grow their own crops. In less than a year, the starvation had ended, and in 1616 the colonists were even able to sell a surplus to the Indians. As settler John Rolfe wrote, "By this means plenty and prosperity dwelleth amongst them, and the fear and danger of famine is clean taken away."

Virginia, however, did not succeed like New England. In New England, land was granted in small plots, making the population independent, property-owning families on small farms, growing a wide selection of crops to feed themselves and become self-sufficient. The land grants were also concentrated in compact small villages, centered on a church where the settler's children could be educated.

In Virginia, the land was handed out to a few wealthy men who had indentured servants do the work for them. Moreover, to increase their power and wealth these landlords devised a system in which they received 50 acres and the forced unpaid labor for a period of four to seven years of each immigrant they brought to Virginia.

Very quickly this small select group of landed plantation owners was buying and selling these workers like cattle. As one servant noted in 1623, "My Master Atkins hath sold me for a 150 pounds sterling like a damn'd slave."

This issue of individual rights must be emphasized. While New England settlers were almost all given rights, land, and status as voting citizens, most Virginia immigrants spent from four to seven years as indentured servants, having almost no rights. Even after their servitude, they were expected to show diffidence to the powerful landowners, and as landless tenants, had no say in how the colony was governed.

With these temporary slaves, Virginia's landowning gentlemen scattered into isolated large plantations. There they tried to squeeze the maximum profit from a single crop economy, devoting as much acreage as possible to tobacco. As John Smith noted in 1617, Jamestown consisted of "but five or six houses, the church down, . . . the bridge in pieces, the well of fresh water spoiled. The store-house they used for the church. The market-place, and the streets, and all other spare places planted with tobacco."

Family considerations

Virginia was grossly different from New England in other ways. Unlike the New England colonies, Virginia was not colonized by families. On the *Mayflower,* the 102 passengers comprised 24 families, including 19 women and 33 children. Later Puritan settlers maintained this policy, immigrating to the New World as men, women, and children.

In fact, family considerations were for most New Englanders a significant reason for immigration. The Pilgrims had fled England partly because they did not wish to raise their children in that society. "Many of their children, by . . . the great licentiousness of youth in that country and the manifold temptations of the place, were drawn away by evil examples into extravagant and dangerous courses," explained Bradford. "[Parents] saw their posterity would be in danger to degenerate and be corrupted."

In Virginia, however, families were never even an issue, and the needs of children were completely irrelevant. The passengers that landed in Virginia on the *Godspeed* in 1607 were all men, including 36 gentlemen, 12 artisans, 14 laborers, and almost a dozen footmen/butlers for the 36 gentlemen. For the next sixty years this pattern never changed, the ratio of immigrants remaining approximately three men for every woman. And the men were generally not heads of families, but single men looking for a "big killing" in a tobacco boom town.

More Than Just Property, Rights, and Freedom

The most significant difference between Virginia and New England, however, were the colonists' reasons for immigrating. While the Pilgrims worked hard to make a profit from their farms, their main reason for immigrating was not money. They came looking for a place to *live.* They brought their families, built farms, and were satisfied if they made enough to raise their children properly, in comfort and security.

More significantly, New Englanders performed these actions under the framework—which for them was religion—of a strong moral commitment. Their robust devotion to right and wrong infused their every action, and was closely linked to how they wished to raise their children.

In Virginia, however, the main goal *was* money. At first the search was for gold, in an effort to duplicate the success of the Spanish. "There was no talk, no hope, nor work, but dig gold, wash gold, refine gold, load gold," said Anas Todkill in Jamestown in the winter of 1607. In later years it was tobacco, grown on every spare plot of ground.

To Virginians, moral commitment and righteous behavior were never a significant consideration. The rights and freedoms of both poorer settlers and black slaves could be easily dismissed if they interfered with maximizing profits. And children didn't even enter into the equation.

The consequence was a place where poverty was rampant, where wealth was concentrated into the hands of a few powerful landowners, and where the rest of the population was poor and struggling. The consequence was a society where children generally grew up ignorant, uneducated, bigoted, and willing to enslave blacks for the sake of wealth and profit. The consequence was a slave state.

Imagine this scenerio for a colony on the Moon. Hardly the vision we conjure up when we fantasize about grand cities in space.

The Formula for Success

Of course, this description covers just two American colonies. Yet, the experiences of the Catholics in Maryland, the French in Canada, the Dutch/British in New York, and the Spanish and Portuguese throughout South America and the Caribbean were all somewhat similar. The evidence from all these colonies says that to create just and humane societies from scratch requires:

1. Stable families, including both men and women.

2. Property rights, in small chunks. In this context, property rights must equate with many individuals or small companies, since the goal is to decentralize power rather than have it centralized under government or a small number of powerful individuals or large companies.

3. Moral commitment, centered on the concept of building a good life for children and the generations to follow.

Working together, these three elements not only built better colonies in the New World, but it also encouraged their growth.

In this context, let now consider the Bush effort to return to the Moon and establish colonies there. Can we glean whether these future lunar colonies will be homes to humane societies, like New England, or will they become our own modern version of a Virginia slave state?

Today's Legal Framework

On September 9, 1997, Jim Benson, Chairman of SpaceDev Corp., announced that his company was going to build the first privately built scientific space probe to travel to another asteroid. "Private companies and the public *can* and *should* have a direct stake in the opportunities space exploration and development have to offer," Benson said. Benson also announced that after the Near Earth Asteroid Probe (NEAP) landed on the asteroid, he intended to "claim" that asteroid as SpaceDev's private property. According to the plans announced in 1997, NEAP was scheduled to launch sometime around January 2000, reaching a target asteroid between nine and fifteen months later.

It is now 2005, and though Benson's company is doing amazingly well, having built the engines used by SpaceShipOne to win the Ansari X Prize, his NEAP project has never happened, and shows little sign of ever occurring. Moreover, even if it eventually does transpire, many will question whether he will be able to make his claim of ownership stick.

In fact, some, including this author, would argue that it is the very discouraging legal environment in space—not technical matters—that has been Benson's main obstacle.

According to the 1967 United Nations Outer Space Treaty, "the exploration and use of outer space . . . shall be the province of all mankind." The Treaty also states that "outer space . . . shall be free for exploration and use by all States." Furthermore, "outer space . . . is not subject to national appropriation by claim of sovereignty, *by means of use or occupation, or by any other means.*" [italics added]

Though most legal and space advocates generally focus on how this treaty has stalled the competition between nations to conquer and acquire territory in space, the treaty's language accomplishes a far more hurtful wrong by denying the right of *any* citizen to own property in space.

To do this, the Outer Space Treaty first prevents the establishment in space of national sovereignty. This denial of sovereignty means that Jim Bensen is forbidden from going to the American courts to settle his claim, since the Outer Space treaty forbids any nation from claiming jurisdiction. Instead, Benson (and all others) must use international law to establish their rights.

Unfortunately, the Outer Space Treaty *is* the international law on this issue, and it is not unreasonable to interpret its language as outlawing such property claims. For example, if Benson was to claim ownership by "homesteading" his asteroid, a traditional and very successful American technique for developing new lands, his action could be construed as appropriation of property by a U.S. citizen "by means of use or occupation" and therefore a national appropiation by the United States, actions expressly forbidden by the Treaty. The Treaty further appears to cement this prohibition by denying the use of "any other means" to claim ownership.

Essentially, the Outer Space Treaty acts to make the ownership of any property in space illegal. While we might have won the Cold War here on Earth, the Soviet Union apparently has won the Cold War in space. The legal framework under which the colonization of space now functions appears remarkable similar to the situation in both Virginia and New England during their first years of settlement (as well as within the Soviet Union under communist rule): a place where everything was held in common and nothing was owned individually.

In such circumstances people starved, and nothing was accomplished.

Not Just Property

Nonetheless, normal human custom states that possession is nine-tenths of the law, and that once any nation or company builds colonies in space, they will be able to claim ownership of that territory. For the sake of argument let us assume that George Bush's space initiative goes exactly as planned, and that sometime around 2015 the United States establishes a permanent base near the south pole of the Moon, taking advantage of a location that happens to have both vast supplies of water ice as well as several mountaintops with perpetual sunlight.

Let us also assume that several of the companies hired by NASA to build that base decide to emulate Jim Benson and SpaceDev and claim ownership of their facilities and land on which they sit. Let us also assume that the international community accepts that claim, thereby establishing the ownership by these companies of this valuable territory on the Moon.

This precedent—that possession deeds ownership—will then become the only guiding principle for determining ownership in space, especially because the Outer Space Treaty forbids nations from implementing more thoughtful legal policy (such as the nineteenth-century homestead policy of the United States).

Simple, crude, and imperfect, this simple land grab will surely stimulate the settlement of space, but it will be a chaotic and anarchist land grab by the largest and most powerful corporations as they rush to claim as much territory as possible. Like Virginia, the control of society will not be spread among many free citizens, but in large blocks owned by a small number of property owners. Like Virginia, the basis for colonization will almost certainly be centered on money and profit, not the building of a new society. And like Virginia, these large corporations will focus on bringing in temporary workers to maximize profits, not families moving to a new world to begin a new life.

Should this scenario occur, based on what happened in Virginia we can expect our future space colonies to be exceedingly unpleasant and harsh places to live.

Moral Purpose

What makes this possibility even more dire is that we—unlike the Puritans—have no deeper moral purpose for going to space, other than our passionate desire to see it. This may very well be the most serious flaw to Bush's present-day space exploration vision.

Few activists talk of establishing human rights for the settlers that will live in these farflung and difficult environments. In fact, the debate usually centers on ways in which we can limit the freedom of future colonists: preventing them from exploiting the resources of space, outlawing their right to use weapons, forbidding them to own property, and generally restricting their ability to live as they wish.

And since the Outer Space Treaty forbids nations like the United States to apply its Constitution and Bill of Rights to the astronauts who will live in these first colonies, we Earthlings have even taken positive action to deny these protections to space dwellers. When astronauts finally arrive on the Moon or Mars to live, they will do so with the understanding that—unlike their brethen on Earth—they have no rights, and are no more than glorified indentured servants. Without the laws that protect each American's freedoms, astronauts will live in a social vacuum—just like the black slaves who were brought to Virginia.

Into this vacuum will flow innumerable other interests. The United Nations will almost certainly try to exert its power, as it attempted to do with the 1979 Moon Treaty, which stated that no part of the Moon "shall become the property of any State, international intergovernmental, or non-government organization, national organization, or *non-governmental entity or of any natural person."* [italics added] The treaty then described how space would instead be administrated by an "international regime" under the auspices of the U.N. Secretary General.

While none of the space-faring nations ever signed that treaty, the General Assembly has repeatedly endorsed its intent, saying for example in 1996 that the exploration and use of outer space should be done "for the benefit and interest of all states, in particular the needs of developing countries." The idea that property in space should belong to those who do the work seems alien to United Nations diplomats.

Other political interests, such as NASA, will also try to monopolize the resources of space. In fact, the NASA bureaucracy today seems to want to claim the entire solar system as its private property, and has even threatened prosecution when it felt its turf was being invaded. For example, grandmother Peg Davis wanted to donate to a museum her father's thirty-year-old pen-and-pencil set in which was embedded a tiny smattering of Moon dust. When NASA learned about the desk set in December 1999, decades after it had been made, several officials were sent to Davis's home to seize it. "[The NASA officials] said they can take it or else they'd seize it with a search warrant," noted appraiser John Reznickoff, who had been asked to assess the desk set's value for Davis.

The Future

W. E. B. Du Bois, in studying the African slave trade, once asked: "How far in a State can a recognized moral wrong safely be compromised?" and answered his own question by saying that it is perilous for "any nation, through carelessness and moral cowardice, [to allow] any social evil to grow. . . . From this we may conclude that it behooves nations as well as men to do things at the very moment when they ought to be done."

Obviously, there are many other more immediate problems that must be solved before future colonies can be built on the Moon, Mars, or even at a space gateway at L1. At the moment we cannot even build spaceships that can go into orbit repeatedly, in a manner similar to the many crossings of the Atlantic made by ships like the *Mayflower*.

Nonetheless, issues of freedom, of moral purpose, of family, of property, must all be considered now, before we have established patterns of social activity that cannot be easily changed.

As a bold first step, the United States should make the blunt and politically incorrect decision to withdraw from the Outer Space Treaty. Article XVI of the Treaty says that "any state party . . . may give notice of its withdrawal from the treaty . . . by written notification." The withdrawal takes effect one year later.

By doing so, we will free ourselves of the Treaty's restrictions. We will permit our own country to set the rules for our own astronauts, guaranteeing that their rights shall be protected under the Constitution.

Once free of the Outer Space Treaty, the United States must then establish land grant policies that allow its explorers to obtain property rights, in reasonable chunks. While Northrop Grumman or Boeing *should* be able to claim territory when these companies build bases on the Moon, it is not unreasonable to place a limit on their claims. Like the American homesteaders of the American West and the Puritan settlers in Massachusetts, small property grants will permit settlement by as many people and companies as possible.

By taking these unilateral actions, the United States might even force the nations of the world to abandon the existing unworkable treaties and work together to write sensible rules for the claiming of territory in space. Rather than making believe that property rights and sovereignty don't exist—as does the Outer Space Treaty—future space colonial powers need an honest and practical framework for establishing their rights and sovereignty.

Knowing that they can claim territory and thus reap the benefits of ownership, many nations will feel compelled to step forward and compete, if only to avoid being left in the dust. Such a competition can only be invigorating.

Choices

Finally, we must honestly ask ourselves: why do we want to explore space? Is it merely "to go where on one has gone before?" By itself this reason, while noble, worthwhile, and truly exciting, is simply not sufficient. We must realize that we also go to expand human civilization beyond the limits of Earth, and that the civilizations we create on the Moon and Mars can either wonderful places to live, or hellholes of tyranny and oppression.

In 1932, Werhner von Braun was faced with that same choice. Passionate in his desire to build rockets and send humanity into space, the private German rocket club that von Braun depended on for research was going bankrupt. He desperately needed cash and resources to continue his work.

Von Braun choose to make a deal with the German army, and hence the Nazis. Like the colonists in Virginia, von Braun let his passion for exploration and rocket-building make his choices for him. He did not think much about the moral dimension of his actions. "I was still a youngster in my early 20's and frankly didn't realize the significance of the changes in political leadership," von Braun wrote years later. "I was too wrapped up in rockets."

Years later, as World War II was ending, von Braun was again faced with a choice: surrender to the Soviet Union or surrender to the United States. Going to the Soviets would be easier. They were closer, and would certainly provide the German engineers with anything they needed to build spaceships.

This time, von Braun took the harder choice, and brought his team to America. He no longer could cooperate with dictators merely so that he could build rockets. As von Braun noted many years later, "As time goes by, I can see even more clearly that it was a moral decision we made [when we chose to come to America.]"

Today, we have a similar choice. We can, like von Braun and the settlers in Virginia, take the quickest and easiest path towards building our colonies in space, and not think about the consequences very much.

Or we can take the harder path, forcing changes in present space policy that might in the short run be difficult and painful, but in the long run bring prosperity and happiness to future colonists.

If we do not take the harder path, however, the future does not look good. Past colonial experience indicates that we are today making the same mistakes that the British made four hundred years ago in Virginia. The consequence there was slavery.

In space it could very well be something as evil, staining those futuristic gleaming cities we now can only dream about.

Robert Zimmerman is an independent space historian. In 1994 he received a Masters degree from New York University in early American colonial history. In addition to his weekly *Space Watch* column for United Press International, he has written three histories about space exploration, *Leaving Earth: Space Stations, Rival Superpowers, and the Quest for Interplanetary Travel* (Joseph Henry Press, 2003), *Genesis, the Story of Apollo 8* (Four Walls Eight Windows, 1998), and *The Chronological Encyclopedia of Discoveries in Space* (Oryx Press, 2000). In 2003 *Leaving Earth* won the Eugene M. Emme Award from the American Astronautical Society as that year's the best space history for the general public.

In recent years, several of us in the pro-settlement community have become proponents of the two-outpost model on the Moon I mentioned earlier. Leading among these are Dr. Wendell Mendell of NASA, Robert (Bob) Richards and myself. However, Bob takes it to the next level. Bob is one of the Founders of the International Space University. That, and being Canadian (a people who seem to be just naturally friendly to everyone) reinforces his ability to step back and look at the opening of the space frontier from an international perspective. His style is always inclusive, not confrontational. When he speaks of the Moon and Mars he does not see a conflict between destinations, but a confluence of needs and opportunities. Putting these two elements together - the international partnership we can forge on the frontier, and the synergies between those who want to go to the Moon and those who want to explore Mars – Bob takes us beyond borders and destinations, to the level of rationales and reason.

LunaMars – The Challenge

by Robert D. Richards

"Nothing on Earth or beyond it is closed to the power of man's reason." - Ayn Rand

As I write this, we are midway between the first humans making landfall on another world and an incredible future; a future that lies ahead waiting to be challenged or ignored, a future that is utterly indifferent to our human destiny. As it has been throughout history, it is up to the explorers and visionaries to define the paths that will lead humanity over the horizon to new worlds. It is up to us: the Apollo-inspired Space Generation. As we look forward from the starting gate of the 21st Century toward the Moon, Mars and beyond, inspired by the heroes of the past and emboldened by the commitments of world governments and entrepreneurs to a new age of space exploration and development, we hold in our hearts and minds the keys to humanity's future among the stars. It is a future bright with astonishing new worlds and perhaps the answers to some of the fundamental questions that have haunted the human psyche for ten thousand years: Are we alone? Is life unique to this pale blue dot we call Earth or are we only the latest chapter in a cosmos alive with a fantastic undiscovered story of universal life? Are we humans capable of becoming something greater than we have been? Are we destined to remain a planet bound species completely defenseless against random or self inflicted global catastrophe? Or can we overcome our brutality to each other and chart a course of peace and prosperity for all human beings through the exploration and development of space?

To become a permanent space-faring species we must begin our journey together by returning to the Moon. As a unifying symbol that spans all cultures and geographies our return to the Moon is a step that will be understood and resonate with every human being. The Moon has beckoned to us since our first ancestors raised a furrowed brow to the heavens. It has watched over us, still and silent, as generation after generation of human beings struggled with their adolescence on a zero sum world. In our time it represents a chance to breakout of a zero sum existence and create new engines of prosperity and hope for every human being; a challenge that will expand the envelopes of our science and technology and inspire great works and efforts by those of us able to transform its lure into tangible and worthwhile goals for the human race.

The New Deal

In 2005 we are at a turning point in human history. Commercial space interests are being driven by a new league of space entrepreneurs. Last year the first private spaceship made an historic flight to prove that space is no longer the dominion of governments. Space is finally being embraced by the private sector, or perhaps more accurately, the conquest of space and all if its promise is being seized by the private sector.

As has been resoundingly pointed out by the X Prize, the most understandable parallels for the development of space are with the aviation age of the early 20th century. Large scale airline transportation systems were made possible with the advent of commercially driven carriers whose frequency of operation drove down the operational costs of air travel to within reach of the individual. This was predated by a perception change within the public – mostly inspired by the entrepreneurial development of aviation technology and in particular by the historic flight of Charles Lindbergh from New York to Paris in 1927.

It is important to remember that the economic development of space will not be realized until commercial business dynamics are in place. Low cost access to space coupled with large scale markets is the key. If you compare today's 747 aircraft with NASA's space shuttle you will see the dramatic difference between commercial and government dynamics. A multi-billion dollar government space truck with ticket prices of more than $20 million per person won't inspire anything except governments. The inversion of logic is clear: instead of hundreds of people watching a handful take off a few times per year, we must get to a point where a handful of people watch hundreds take off thousands of times per year. In this century the commercial space industry will generate economic demands for space technologies and infrastructure analogous to the supply chain supporting today's aviation industry. Any company with key space technologies will be able to set up manufacturing supply lines based on recurring costs and long term profit margins.

Will this happen within a five-year return on investment constraint? No, not likely. But this is the long term market landscape: unlimited commercial space opportunity fueling tourism, exploration, resource development and eventually human settlement, wherein the cycle begins again. The opportunities for wealth creation are vast. As space entrepreneur Jim Benson has said, "Who wants to be a trillionaire?"

Never before have we seen such private sector buy-in to the space frontier. There is a new breed of space entrepreneur emerging. Wealthy visionary captains of industry, perhaps best described as "venture philanthropists", are plotting private sector commercial ventures that will achieve far greater things far sooner than any of the programs contemplated by the world's space agencies.

However, the role of government is an important one that if properly channeled can become catalytic to the exothermic reaction we need for the true commercialization of space. As we embark on the new millennium, many nations have expressed a renewed vision for exploring our neighboring worlds and creating a permanent human presence beyond Earth. Space is being increasingly used as a platform from which to understand, protect and secure our home planet. A new space renaissance is emerging.

After decades of moribund vision, NASA is now looking at space exploration and its role in the true context of economic development and human history. This relatively sudden shift in thinking at the top is taking much of its management inertia by surprise, with vast repercussions both internally and to its international space agency partners. When a character of an organization as large as NASA changes; a number of internal and external crises are sure to erupt. However it's when crises collide that the true character of an individual or organization is demonstrated. And the same can be said for human society as a whole. NASA's challenge to reform is a microcosmic reflection of the challenge facing the entire human race.

Charting The Course

"Some day, I know we will prevail on Mars and the other planets, putting forth something of ourselves and shouting, "I'll be damned! Yes! We had a vision. Now we know it was right!""

- Ray Bradbury

We are the new explorers and dreamers who are responsible for laying the framework for this grand new adventure. We must cultivate the ideas and inputs of scientists, engineers, policy makers and entrepreneurs from many nations and put these ideas into motion with concrete plans and roadmaps. Together we must work hard to chart the course of space exploration and development over the coming decades.

To open the space frontier for the ongoing benefit of all humanity, we must embrace the many aspects of our human civilization that drive our political and social landscapes. In particular we must expand our mindset to embrace the multitude of cultures and societal imperatives that drive the value equations of our existence. Science for the sake of knowledge alone will not satisfy this. Technology advancement without social application will not satisfy this. Feats of exploration bereft of economic sanity and sustainable development will not satisfy this. To expand outward as a mature space faring species, humanity must do so by agreeing on a challenging but sustainable and socially responsible common international framework for the exploration and development of our neighboring worlds.

LunaMars

The Moon is not just a stepping stone to Mars, nor is it a destination that we should go to now and worry about Mars later. Humanity needs to explore and develop both these worlds. In doing so Mars and the Moon require similar core space capabilities while providing very different opportunities for humankind that stand on their own merit.

In our treasure chest we have the vision and entrepreneurial drive of the United States; the collaborative strengths of the European Union; the enduring space heritage of Russia; the sophisticated niche technologies of Canada; the motivation and innovations of Japan; the long term commitment of China; and the remarkable efforts of the developing space nations exemplified by India. The synthesis of all of these human capabilities and desires can result in a permanent human presence on both the Moon and Mars in the early part of this century. And, to paraphrase the great writer Robert Heinlein, from there we are halfway to anywhere.

The differentiation between the Mars versus Moon advocacy communities is largely based on the truth that Mars is not like the Moon and vice versa. The argument that learning to live on the Moon will have direct relevance to the requirements of living on Mars is not without holes. We cannot argue that learning to live in a settlement of the Australian outback is relevant to learning to live in an Antarctic outpost just because they are both far away. There are some common lessons to be learned but experience in one environment is not sufficient to ensure survival in the other. However by evoking a synthesized approach to space exploration with the Moon and Mars as essential stepping stones toward a larger vision, we may be able to galvanize the professional communities as well as the public and private sector into a plan that makes sense for everyone.

LunaMars

The LunaMars concept is based on the establishment of an international framework in which the exploration and development of the Moon will enable the core competencies and experience that can be most effectively applied to the exploration and development of Mars. This international framework is designed to galvanize the Moon-Mars communities while establishing clear roles and goals for both robotic and human space exploration. To accomplish this, it is necessary, but not enough, to establish activities and human presence on the Lunar Nearside. Nearside infrastructure has its own scientific and technological merit and is certainly the next logical step in our return to the Moon. But before committing humans to go to Mars there is a strong rationale to first establish a Mars analog base on the Lunar Farside – LunaMars - where many technical, psychological and extreme environmental conditions mimic those that will challenge future Mars operations and human presence.

The primary objective of a LunaMars base on will be the ability to truly isolate humans from their home planet and practice living without an umbilical cord. The technical, psychological and sociological factors of true human isolation from Earth have never been simultaneously tested. So the first purpose of the LunaMars base is a test bed for how we humans might react to leaving the nest for the first time. Throughout history pioneers have set themselves off on voyages that took them far from home with varying levels of postpartum trauma but never to the extent of what will be required for a Mars mission. Earth's explorers and settlers are the most relevant examples we have to judge what will happen in the minds of those who watch their familiar home fade into unknown territory, and modern day astronauts have caught a glimpse of this as they looked upon Earth from orbit or even from the Moon. But no human has ever been committed to a long duration separation from the rest of humanity and their home planet.

There is a duty to responsibility here. We have seen what occurs in the aftermath of tragedy when the public perceives great risk in space exploration. The first thing that happens is everything stops. The system shock is too great. And then investigative hindsight reveals the internal failures of the program in question, as we witnessed in the two shuttle disasters. Each time we discovered design and management errors that could have been avoided. Notwithstanding the terrible personal tragedy involved, these exercises bring out the inherent nature of government space programs: the inversion of purpose as dictated by the perceived public mindset. There arises therefore a complex interplay of public emotion and fiscal outrage as it applies to the loss, and cause and effect become entangled and sometimes indistinguishable. Large groups by nature are risk adverse and publicly funded space programs will never outpace the risk required for pioneering accomplishment. We have learned that the human psyche a public space program evokes cannot absorb the manifest risk involved. It is therefore tantamount to disaster to stage a publicly funded Mars mission without the due diligence required to ensure that every aspect of the mission has been tested and practiced – including the public/private models to ensure sustainability and economic sanity as well as safety.

The metaphors closer to our every day lives are common. It's like the young children who approach their parents with the announcement that they want to go camping. Okay, the parents respond, but first you are going to practice in the backyard. So the kids set up their tent and get all they think they need in order and happily plunge themselves into their make believe remote and independent world. Then they realize they forgot to go to the bathroom. And perhaps those extra batteries mom suggested weren't such a bad idea. And they learn all the things we knew they would learn but within safe haven. Such is the necessity before we commit humans to go to Mars. Yes we can practice on Mars analog outposts here on Earth; yes we can practice technologies and techniques in orbit and even on first lunar robotic and human outposts on the lunar nearside. But before we commit to that next gargantuan and irrevocable step; before we send our children beyond the reach of safe haven, we must first synthesize all we know in an environment as close to Mars as possible but still within quick reach of the cavalry. The only place we can do this is on the Lunar Farside. If we don't do this; if we don't create LunaMars and mitigate every risk within our means to do so, if something goes terribly wrong with the first human Mars mission and people die, the magnitude of the system shock will overshadow anything we've ever experienced in the space program to date - and we may not recover for centuries.

The good news is that we have a natural extraterrestrial lab where we can do our due diligence. A number of things would happen in the LunaMars outpost that can't be convincingly achieved anywhere else. Perhaps most significantly, our LunaMars pioneers would be visually cut off from any sign of Earth. Because of the moon's gravity-locked synchronous orbit the Earth is never visible from the Farside. We have absolutely no idea how the human psyche will react to this. The second isolation aspect of the LunaMars outpost is communications. Direct communications with the Earth would not be possible. Also, the Farside is completely shielded from Earth with no hint of the cacophony emanating from our noisy planet. It is absolutely silent. No TV, no radio, not even CNN; nothing except what comes from the stars. It is humanity's largest natural isolation lab. We have the opportunity therefore to introduce controlled communications to mimic what Mars-bound humans would experience. A satellite relay in lunar orbit could be outfitted with a delay to replicate the Earth-Moon transit time.

We can set up LunaMars in exactly the same way we would prepare for the first human Mars mission. First the reconnaissance and site selection, then the robotic precursors and eventually the humans arrive. It is in this very first stage, the exploration, that many different nations can, and in fact, already are, contributing in a high profile way. Several different probes, orbiters and landers are on the drawing boards around the world, and some are on their way to the Moon right now. By establishing a coordinated approach to this first wnew wave of exploration, dubplication can be reduced and knowledge expanded at the lowest possible cost. It also will set a precedent of international sharing of the risk and opportunities,while allowing the inspiration of space exploration to be shared in a direct way around the globe. After all, a small nation can't afford to send its own human crews to the Moon, but it is well within its grasp to build and fly a lunar comsat.

Once there the LunaMars pioneers begin there work in living and learning how to survive independently. They will need to deal with radiation and all its dire implications and consequences to humans and machines. They will become their own experiments in physiology as the long term affects of low gravity can be traced over time and correlated with existing knowledge. They will test and implement sustainable and safe nuclear power generation without which a Mars colony would not survive. And perhaps most importantly in the context of historical analogs, they will learn to live off the land, developing and adapting materials using in situ resource utilization while perfecting off-Earth agriculture and closed habitat systems. When the necessities of life are in hand our LunaMars pioneers can begin efforts to do science and put into practice models of economic trade with Earth. Knowledge will be one of the principal LunaMars commodities and will be gained through taking advantage of the unique isolated environment where new research efforts in astronomy and biology are made possible.

How to we get to LunaMars? We get to it by continuing the robotic reconnaissance of the Moon and Mars, and while doing so setting up a framework of four primary steps for a safe and enduring human presence beyond Earth:

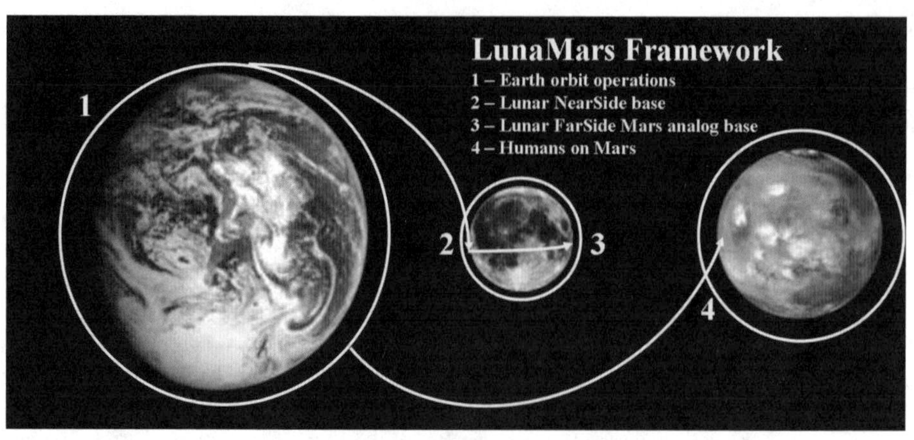

1 – Earth orbit operations – core competencies via ISS + other infrastructure
2 – Lunar NearSide base – core competencies in robotic/human planetary exploration
3 – Lunar FarSide Mars analog base – international isolated robotic/human outpost
4 – Humans on Mars – apply LunaMars experience to the challenge of Humans on Mars

There is a vital concept interweavoven within all of this that must not be overlooked: the economic model adopted now for Moon and Mars exploration and development will dictate the success or failure of our species to live permanently on other worlds. A government-only conceived, implemented and operated model will never achieve the exothermic reaction needed for economic sustainability. It is absolutely essential that the government bridge the gap of exploration without becoming its own customer. This is particularly relevant for NASA, who to date has done the most in space at the the most possible cost. The cold war Apollo model worked to achieve flags and footprints but little more. The shuttle and even the ISS program models did not have commercial economics in mind when conceived. The X Prize has shown that the private sector can do wondrous things when met with a challenge and left alone to achieve it. We will need a new way of thinking bred into our new model that engages a hybrid of public and private sector interests. Everything that is built now must have commercial succession in mind. Otherwise only a few will go and scratch rocks while the rest of the market stays at home and watches with rapidly declining interest.

The LunaMars canvass is blank and will be filled in over time by the community of government, public and private sector stakeholders in all of their diversity. The overriding objective of the LunaMars concept is to define a common international approach to establishing an enduring human presence beyond Earth and in doing so avoiding unnecessary conflict, duplication and risk. The goal is to deliver a framework that has teeth; a framework with clear goals and roles within an overarching rationale that can eventually be supported scientifically, technically and politically by all international stakeholders.

Making It Happen

Yes we are at a turning point in human history; but a turning point can also be a precipice. If we do not meet the demands of our growing planetary culture in this century we could see humanity spiral into global chaos and war, a tail spin from which civilization as we know it may never recover. It is our challenge at the dawn of the 21st century to catalyze and consolidate the allies of humanity and reason and reach a new world order of peaceful cohabitation and development. Not without pain, not without trauma, but with new planetary antibodies that disallow the cancerous plagues of hate and violence within our society combined with societal endorphins that produce the freedom and harmony required for the happiness and prosperity of every human being.

Brilliance and Leadership. Inspired Dreams. Strategic Thought. Friendship and Collaboration. These are the engines of cultural and economic change and the hallmarks of the greatest human adventures. We now have the technology to invoke these engines of change on a planetary scale. This is the Challenge to each and every one of us imbued with evolutionary foresight: to turn our Vision into Reality and make amazing things happen.

Robert (Bob) Richards is Director of the Space and Atmospheric Division at Optech Inc., a world leading developer of advanced laser radar (lidar) systems for space operations, Earth observation and planetary exploration. Bob studied space science at Cornell University, where he became special assistant to Carl Sagan. In 1987 he co-founded the International Space University (ISU), where he served as the university's first associate administrator for strategic planning and chaired the board's administrative and strategic planning committees. Prior to ISU he co-founded the Space Generation Foundation, whose youth outreach programs include Yuri's Night and the Space Generation Congress, with UN Observer Status and space policy activities through its sister organization, the Space Generation Advisory Coucil. As a student Bob co-founded Students for the Exploration and Development of Space (SEDS), which continues today as the largest student run space organization in the world. He is a contributing author of "Blueprint for Space", published by the Smithsonian Institution.

Some may think that opening a frontier such as space is simply a matter of having the right technology, getting the funding, and "hitting the road." That's far from it. We are rapidly reaching the point where governments and private citizens will have routine access to space. This means we will soon not only be traveling through this new domain, but many will want to become active there. From building habitats, to developing new products and services, to harvesting resources such as those we will find on the Moon, all sorts of activities will begin to occur. On Earth we take these for granted, as we have had thousands of years to develop the legal systems needed to protect those that create new wealth, and those that might be harmed if these activities go awry. But not so in space. Its legal system is literally a blank slate. So how do we protect inventions developed in space? Who enforces the laws? What do we do to protect the space environment? And how do we set up systems that encourage human expansion, rather than inhibit it? Rosanna Sattler lays out a comprehensive framework on the current state of law which might apply to these questions. Her essay is required reading for anyone who wants to understand the matrix (and morass) of space law.

Transporting A Legal System From The Earth To The Moon

by Rosanna Sattler, Esq.

Introduction

By the year 2020, humans will return to the Moon and do much more than simply explore its surface. It is anticipated that an infrastructure will be developed to encourage space tourism and commercial operations. The effective use and management of the Moon and its resources are integral to the economic development of space and the expansion of business and industrial enterprise there.

In January 2004, President George W. Bush announced his vision for the future of the exploration of space and the extension of the human presence across the solar system. This vision is infused with a "renewed spirit of discovery" aimed at exploring the Moon, Mars and beyond. It also encourages the development of space resources and infrastructure. One of the goals is living and working on the Moon for increasingly extended periods of time. Later that month he created the President's Commission on Implementation of United States Exploration Policy ("the Commission.") This Commission held public hearings and heard testimony from individuals in industry, education, media and various agencies and professional organizations on ways to expand space exploration, discovery and commercialization by private entities. In June 2004 the Commission published numerous recommendations in its final report, including ways to streamline and reorganize NASA. These recommendations called for greater reliance on private industry in space operations, reducing NASA's involvement to "only those areas where there is irrefutable demonstration that only government can perform the proposed activity." The Commission also addressed ways in which the government could engage private industry and expand commercial involvement in space.

Specifically, the Commission recommended:

...that Congress increase the potential for commercial opportunities related to the national space exploration vision, by providing incentives for entrepreneurial investment in space, by creating significant monetary prizes for the accomplishment of space missions and/or technology developments, and by assuring property rights for those who seek to develop space resources and infrastructure.

Space holds the promise of vast new opportunities and untapped resources. The Commission recommends that the United States encourage and accelerate the economic development of space. The Commission report lists several incentives that will likely entice private industry to invest their resources and capital in space ventures. Recommendation 5-2 by the Commission recommends that Congress increase the potential for commercial opportunities related to the national space exploration vision by:

Providing incentives for entrepreneurial investment in space. Creating significant monetary prizes for the accomplishment of space missions and/or technology developments Assuring appropriate property rights for those who seek to develop space resources and infrastructure

The Commission suggests creating a $100 million to $1 billion prize to be offered to the first private entity to place and sustain humans on the Moon for a specified period of time. One incentive has already been shown to spark "entrepreneurial investment" in space technologies. In October 2004, the non-profit X-Prize Foundation awarded a $10 million Ansari X-Prize to the spacecraft SpaceShipOne, for achieving suborbital flight twice within one week. The report estimates that over $400 million was invested by competitors in developing their technologies, a 40-to-1 payoff reward for the development of this technology. Corporate sponsors, including M&M Candies, paid an estimated $2 million to have their logos displayed on SpaceShipOne. Richard Branson, CEO of the Virgin Group, which includes Virgin Airlines and Virgin Records, reportedly agreed to pay up to $21 million over the next 15 years to provide spaceships and technology for a proposed sub-orbital space airline, Virgin Galactic. Discussions are underway for similar deals with four other spaceline operators. A director of Virgin Galactic states that the company is prepared to invest another $100 million to build larger vehicles to carry paying customers. The first five-passenger flights are planned for 2008, with ticket prices set at $210,000. Approximately 18,000 people from around the globe have made reservations for the flights.

The birth of this nascent commercial space tourism industry is supported by President Bush, who on December 23, 2004 signed into law The Commercial Space Launch Amendments Act of 2004. This law also eliminates confusion over what government agency should regulate sub-orbital aircraft, placing authority under the Federal Aviation Administration ("FAA") Office of Commercial Space Transportation. Under this law, the FAA will regulate the industry over the next eight years primarily to protect the uninvolved public and the public interest. However, the FAA will regulate space vehicles to ensure crew and passenger safety, only if the actual operation of those vehicles results in death, serious injury or a dangerous close call. Beginning in 2012, the FAA can regulate spaceships however it sees fit. The eight-year period will give spaceship developers more freedom to experiment and also allow them to generate revenue by taking on passengers, as long as they know the risk.

In 2004, NASA created the Centennial challenges prize program which provides up to $10 million each year for technological advances. Supporters hope that these prizes will stimulate private financing of missions. Other private entities are offering large cash prizes for specific accomplishments as well. In November 2004, Bigelow Aerospace announced its "America's Space Prize," which will award a $50 million prize to a United States-based contestant who builds a spacecraft that can carry a crew and dock with an inflatable space habitat developed by Bigelow Aerospace by the January 10, 2010 deadline. NASA also has a prize program, the Centennial Challenge, which provides up to $10 million each year in cash prizes for technological advancements. However, the upper limit of each prize is $250,000, unless there is special legislative authority for an increased limit.

The Commission report also promotes the creation of tax incentives for private industry, such as making profits tax-free until they equal five times the original investment, or tying tax incentives to specific milestone achievements. Additionally, it cautions against the over-regulation of this burgeoning industry through overly restrictive occupational safety or environmental regulations.

In addition to financial incentives, the report recommends protecting and securing the property rights of private industry in space. However, the report offers little specific direction as to how property rights in space are to be created and protected, though it does point out that two treaties exist, the Outer Space Treaty and the Moon Treaty, that may make such ownership difficult. In fact, the report states:

Because of this treaty regime, the legal status of a hypothetical private company engaged in making products from space resources is uncertain. Potentially, this uncertainty could strangle a nascent space-based industry in its cradle; no company will invest millions of dollars in developing a product to which their legal claim is uncertain. The issue of private property rights in space is a complex one involving national and international issues. However, it is imperative that these issues be recognized and addressed at an early stage in the implementation of the vision, otherwise there will be little significant private sector activity associated with the development of space resources, one of our key goals.

The implementation of the President's vision requires an overhaul of the current treaties and laws that govern property rights in space, in order to develop better and more workable models that will stimulate commercial enterprise on the Moon, asteroids and Mars. The expansion of a commercial space sector to include activities on celestial bodies requires the establishment of a regulatory regime designed to enable, not inhibit, new space activity. The development of specific laws, which are consistently applied, will create a reliable legal system for entrepreneurs, companies and investors. The establishment of a reliable property rights regime will remove impediments to business activities on these bodies, and inspire the commercial confidence necessary to attract the enormous investments needed for tourism, settlement, construction and business development, and for the extraction and utilization of resources.

EXISTING INTERNATIONAL SPACE TREATIES ARE UNWORKABLE FOR PROPERTY RIGHTS

Currently, there are several treaties in effect that were created to address space exploration. Most of these treaties were drafted during the 1960s "Cold War," when outer space was seen as the next battlefield, and the Moon as a potential military outpost. These fears were fueled by the "space race" between the United States and the Soviet Union, each trying to beat the other to the Moon. Other nations feared that the two rising superpowers would dominate space and claim it for themselves. In 1967, in response to these fears, the United Nations drafted the Treaty on Principles Governing the Activities of States in the Exploration and Use of Outer Space Including the Moon and Other Celestial Bodies ("Outer Space Treaty.")

The Outer Space Treaty provides that space exploration is to be for the benefit of all nations, that there shall be "free access to all areas of celestial bodies" for all nations, that exploration must be "in accordance with international law" and that "the Moon and other celestial bodies shall be used... exclusively for peaceful purposes." The Treaty also contains several beneficial provisions, such as a requirement that states must "render...all possible assistance in the event of accident, distress or emergency..." to astronauts, who are to be viewed as "envoys of mankind in outer space."

The provision of the Outer Space Treaty which has caused the greatest controversy and discussion is found in Article II: "outer space, including the Moon and other celestial bodies, is not subject to national appropriation by claim of sovereignty, by means of use or occupation or by any other means." There is disagreement about whether this treaty restricts the ability of individuals to hold property rights, or whether it simply restricts the rights of sovereign nations to claim portions of celestial bodies. Some commentators argue that the restrictions placed on sovereign nations are extended to individuals through their citizenship, and therefore individuals and individual companies may not claim property rights in outer space. There is also disagreement as to what appropriation is prohibited. Some argue that the appropriation clause simply bars ownership of the land, not the resources found within the land, which can be extracted and removed as private property. Others argue that the resources are part and parcel of the land and cannot be treated separately from it. The appropriation provision of the treaty is arguably unclear and undefined and therefore unworkable. Critics argue that the provision is a result of the socialist ideals that were prevalent at the time but it is outdated and at odds with today's prevailing free market economy.

Several international agreements were enacted that expand specific concepts and language found in the Outer Space Treaty, and provide helpful principles which can be revised and clarified in the development of a cohesive legal system applicable to activities on the Moon, Mars and asteroids. In 1968, the Agreement on the Rescue of Astronauts, the Return of Astronauts and Objects Launched into Outer Space was signed. This agreement requires those nations involved in space operations to conduct rescue operations and return astronauts and spacecraft to the appropriate country. This broad mandate is not accompanied by any specific guidelines, and is silent on such matters as which country would retain the financial obligation for such an operation. The Convention on International Liability for Damage Caused by Space Objects was created in 1972 to address questions of financial responsibility, in the event that a spacecraft or other object damages other space objects, the Earth or other aircraft. Though this Convention does provide a mechanism for dispute resolution through the United Nations, it is limited in application to instances in which both parties are members of the UN. In 1975, the Convention on Registration of Objects Launched into Outer Space mandated that each state maintain a detailed record of all objects launched into space and that this record be provided to the UN, further streamlining the determination of liability should the object cause damage in space.

While these treaties further clarified certain provisions of the Outer Space Treaty, many ambiguities remained as space technology continued to advance in the 1970s. A second major treaty, the Agreement Governing the Activities of States on the Moon and Other Celestial Bodies ("Moon Treaty") was signed in 1979, as the expanding US space program led to the possibility of actually using lunar resources. This Treaty was not widely accepted, and no major space power has signed it because it further restricts ownership and prohibits any property rights until an international body is created. The Moon Treaty does allow "states parties...in the course of scientific investigations to use mineral and other substances of the Moon in quantities appropriate for the support of their mission," and it permits individual states to construct space stations on the Moon and retain jurisdiction and control over these stations. The Outer Space Treaty provides that all stations, installations, equipment and space vehicles on the Moon and other celestial bodies shall be open to other States Parties to the Treaty on the basis of reciprocity. Representatives must give reasonable advance notice of a projected visit so that appropriate consultations may be held and that maximum safety precautions may be taken to assure safety and avoid interference with normal operations in the facility to be visited.

Both the Outer Space Treaty and the Moon Treaty have proven to be unworkable foundations for the creation of a usable property rights regime in space, given their ambiguity, lack of support and the controversies surrounding their Cold-War influenced provisions. However,

The Moon
(NASA)
She hangs over us each night, beckoning our return. The Moon. It has been said that if God wanted us to leave this planet, we would have been given a Moon. And we were. Like a nearby offshore island, it is the obvious first place for our civilization to reach, and for some of us, someday to call Home.

Apollo 17
(NASA)
Jack Schmitt and Gene Cernan were the last two human beings to walk on the Moon. Ironically, Schmitt was also the first scientist to go there. What he and others have found tell us that the Moon has what we need to survive when we return. Unfortunately, today's generation sees these pictures the same way they do those of the World Wars - something great that happened in the distant past.

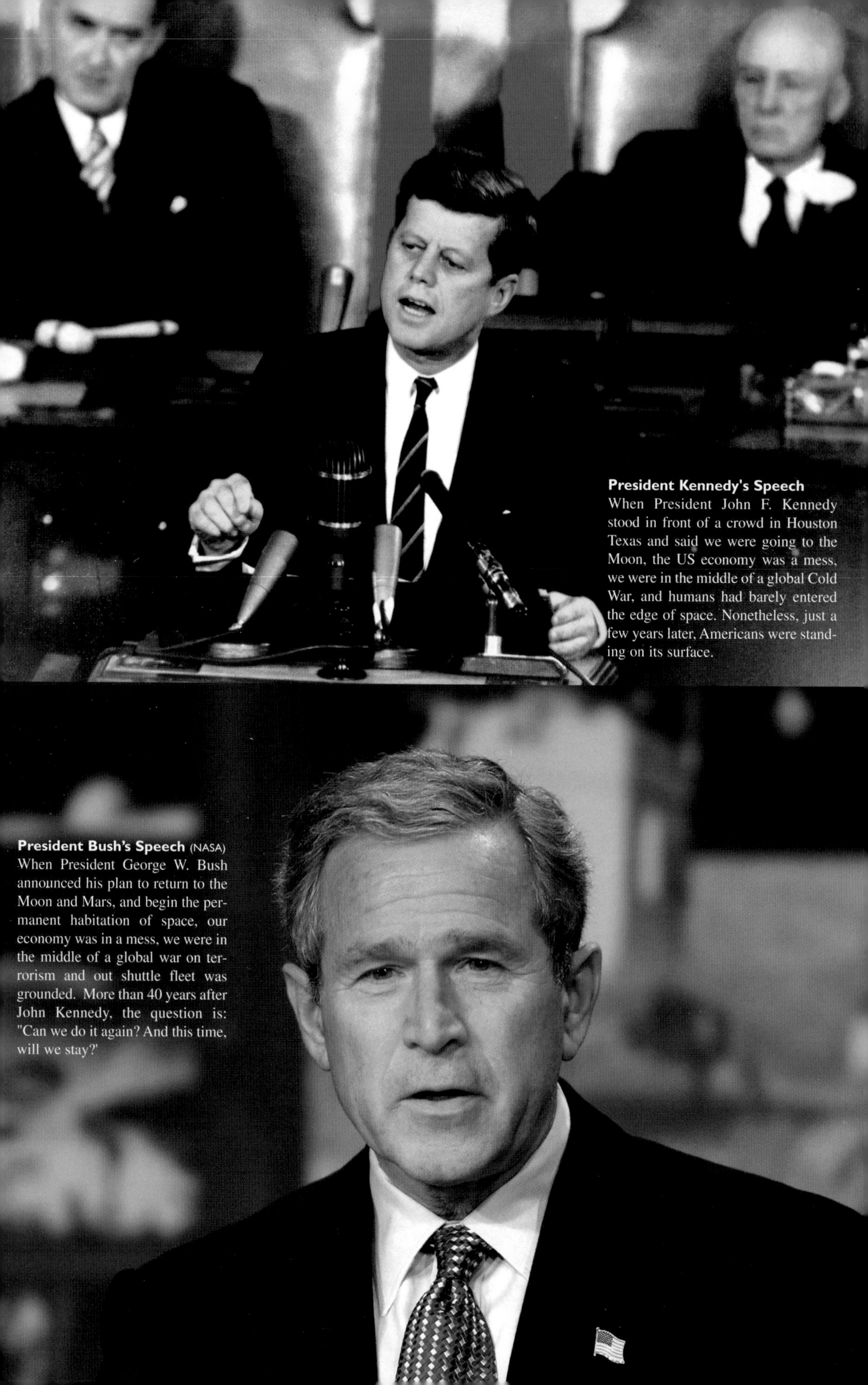

President Kennedy's Speech
When President John F. Kennedy stood in front of a crowd in Houston Texas and said we were going to the Moon, the US economy was a mess, we were in the middle of a global Cold War, and humans had barely entered the edge of space. Nonetheless, just a few years later, Americans were standing on its surface.

President Bush's Speech (NASA)
When President George W. Bush announced his plan to return to the Moon and Mars, and begin the permanent habitation of space, our economy was in a mess, we were in the middle of a global war on terrorism and out shuttle fleet was grounded. More than 40 years after John Kennedy, the question is: "Can we do it again? And this time, will we stay?'

Neighborhood (NASA/Pat Rawlings)
Two key frontier concepts that apply to our return to the Moon are modularity and building up resources from the first missions. Here you see a cluster of modules, all based on a similar design (the tall tubes are solar collectors/ radiators). Many of these would have been launched to the surface before humans arrive. The vehicle is based on the NASA Crew Exploration Vehicle in vogue during 2005. Its aerodynamic shape comes from its ability to fly to the Moon and return to the Earth's surface - to some, a controversial approach.

Cold Storage (NASA/Pat Rawlings)
Although the first new lunar explorers will be robots, soon after we will want to have humans on the scene. Like the wheel on this rover, sometimes things break, and there is no substitute for a human touch! We will also see combined missions and activities, where NASA and other government entities work hand-in-hand with commercial firms. These two teams are working on the rim of a South Pole crater, at the bottom of which may lie millions of tons of ice - lunar gold!

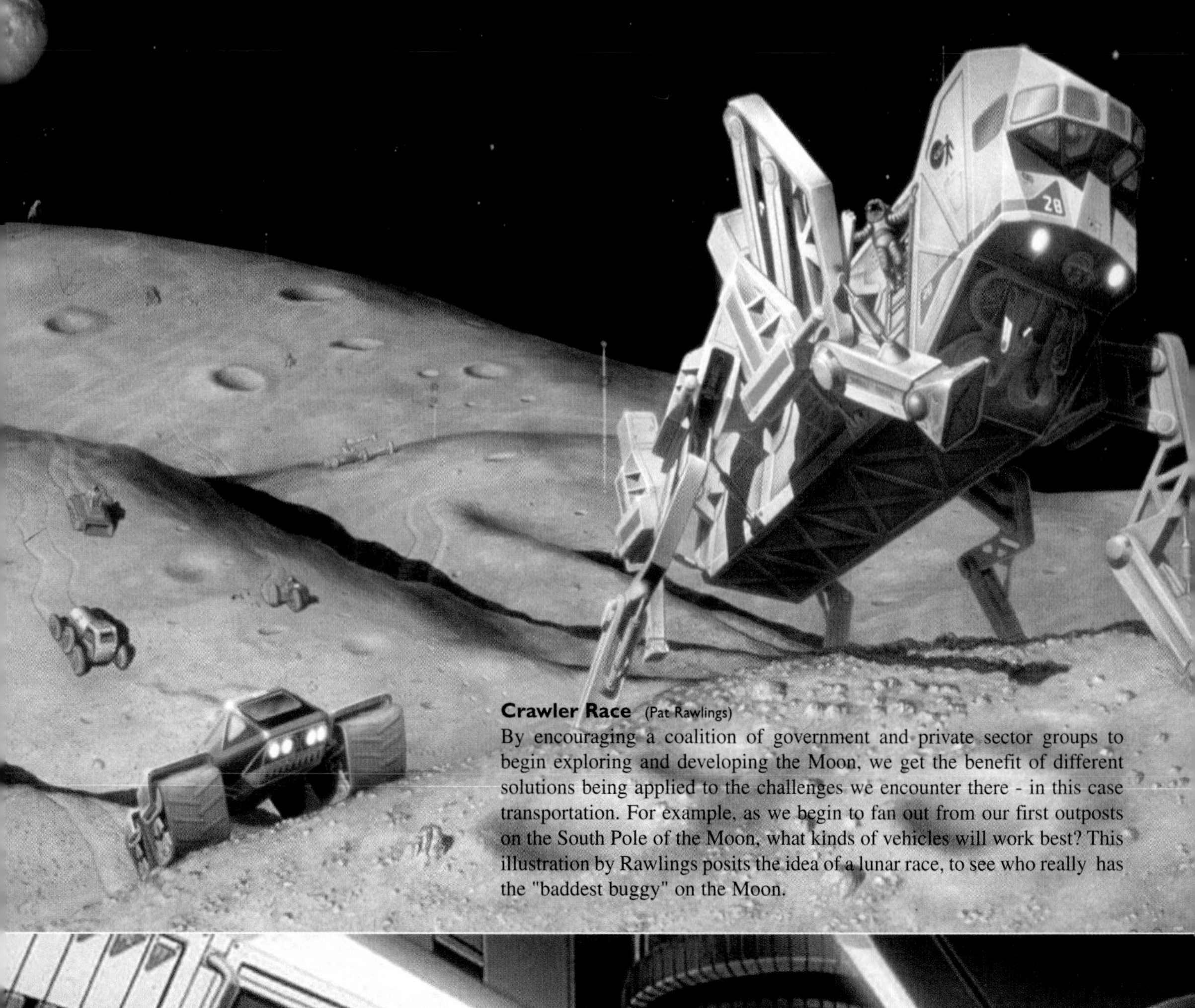

Crawler Race (Pat Rawlings)
By encouraging a coalition of government and private sector groups to begin exploring and developing the Moon, we get the benefit of different solutions being applied to the challenges we encounter there - in this case transportation. For example, as we begin to fan out from our first outposts on the South Pole of the Moon, what kinds of vehicles will work best? This illustration by Rawlings posits the idea of a lunar race, to see who really has the "baddest buggy" on the Moon.

Field Trip (NASA/Pat Rawlings)
One of the greatest differences between the opening of the space frontier then and now is that for the first time we are all going together. A united humanity will be moving outward, and those who choose to go will only be limited by their courage, imagination and drive. It won't matter what part of Earth you came from, what color or gender you are, as long as you can be counted on to be a part of the team - for everyones' survival in space depends on everyone else. Perhaps this is a lesson we could use back here on Earth.

911 (NASA/Pat Rawlings)

Exploring space is not a video game; it's real. And in reality people can get hurt, or worse. The Moon provides us a place to begin to learn how to survive and how to handle crises. In this case it is only a broken leg, which is being tended by medics who have just arrived on a Lunar hopper, and are using heads-up displays to talk with home base. If the injury is bad enough, or the need great enough, the patient can be sent back to Earth in a few days, or medicines shipped out to them. When we make the trip to Mars, however, it will be months before help can be had (so we'd better know what we are doing by then.)

Lunar Miner (Pat Rawlings)

The resources available to us on the Moon are immense. From precious metals to helium3 for use in medicine and energy production, to the elements to sustain life such as oxygen and hydrogen (which are also the key components of rocket fuel), it is simply a matter of patiently filtering and processing enough Lunar soil to get what we need. It is also important to remember that unlike the Earth, we will be disturbing no living ecosystems or fragile habitats, just lifeless rock.

Vigil (Pat Rawlings)
There are a myriad of activities we can do on the Moon, which when combined add up to an economic and scientific basis for our return. Among them are building giant astronomical observatories, such as this one on the Moon's far side. By spacing several telescopes out over a wide distance, and using computers to combine their images, we can create an image as if we had a telescope the size of the entire array. These "interferometers" can help find continents on other Earth-like worlds, and look back to the beginning of time. Unlike telescopes floating in space, these will need no rockets to keep them aligned, and can be serviced by someone walking up to them.

Homestretch (NASA/Pat Rawlings)
In early missions it will be important to maximize the use of each component sent to the Lunar surface, while minimizing the chance of complex systems breaking down. There are several ways to deal with this challenge. Extending the modular concept to habitats, this image shows a cylindrical module which sits in a wheeled cradle for traverses. This approach has the advantage of minimizing Extra Vehicular Activities (EVAs), and the complex interfaces such as tunnels etc., while keeping exposure to Lunar dust to a minimum.

Conversion (NASA/Pat Rawlings)
A similar "LunaMod" vehicle to the previous image pulls up and under a lunar space "tug" or carry-all. Again, the system is kept simple, and the number of elements kept low. The top of the vehicle carries different payloads and additional life support supplies.

Village (NASA/Pat Rawlings)

As the outpost grows, the systems will be refined, and those that work, repeated. Commercial and research activities will co-exist, with personnel from both sharing facilities and infrastructure. The economics of scale will begin to come into play, as the second landing pad will be cheaper than the first, and so on. By spreading the costs between the government and private sectors, each will benefit.

Moonville (Pat Rawlings)
Eventually we will require ever larger facilities. Using inflatable technologies, giant domes can be constructed to house almost the entire community, eliminating the need to go outside to get from point to point. This is what one such hab might look like before being covered with Lunar soil to protect it from cosmic radiation.

Plaza (NASA/Pat Rawlings)
Given its low gravity, it is not hard to imagine large-scale structures being built on the Moon. There will come a day when the first trees are planted there. Green central areas such as parks and commons can be the center of life and connect us to the biosphere back on Earth. As this image illustrates, in such a large hab, pressurized to one Earth atmosphere, a human being can achieve the greatest of all dreams: to strap on a pair of wings and fly!

Leap of Faith (NASA/Pat Rawlings)
Eventually the ability to travel to and from the Moon may mean the development of not just a fairly large permanent population, but a viable space tourism industry, complete with tourists and visitors. Among the attractions will be activities based on the Moon's low gravity of one-sixth that of Earth. Here, a lot of incredible things can happen in the world of athletics. Every jump, pole vault and throw can be six times farther, longer or higher. Supplemented with human aeronautics, the Lunar Games would be quite an event.

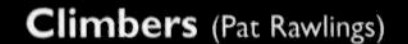

Climbers (Pat Rawlings)

High above what is now most definitely a city, adventurers and explorers try their hand at climbing. The ability to experience the new and unknown for yourself, rather than through appointed government explorers or the cameras on robots, is a core principle of the frontier movement. Being out there, taking the risks and reaping the rewards, be they spiritual or economic, is a fundamental right, one that with any luck you and your children will get to experience someday on the Moon.

Bottomland (NASA/Pat Rawlings)

Here we see a domed city at the bottom of a shaded crater on the South Pole of the Moon. Using giant reflectors to bounce sunlight, a day/night cycle can be created, perhaps matched to a time-zone back on Earth. It seems fantastic, but with determination and the right decisions now, this could happen. After all, many a modern U.S. city started out as a simple outpost or fort on an empty and apparently useless wilderness. If we let the people of Earth do what they do best, which is create, invent and transform the unknown to the known, anything is possible. Then, like tiny bubbles of life, human habitats will eventually spring up on the Moon and throughout the solar system.

With NASA out of the Earth to LEO business, new players can develop different vehicles to compete for different markets. Using the current government or a new commercial space stations as transfer points, travellers can then board more specialized craft for the longer trip to the Moon. Thus, we build up a strong "Space Port" infrastructure, with commercial and government facilities such as hotels, labs and fuel depots to support outposts on the Moon and Mars. SKYCORP/Mark Maxwell

The development of a robust Earth/LEO/Moon infrastructure is the key to opening the frontier of space. With the ability to build and operate fleets of various craft as they carry people and payloads from one destination to another, will come the expertise and industrial culture which can then be used to open Mars and eventually the rest of space. Just like on Earth, in space it is all about location, location, location! For example, this image shows an L1 space station, which sits at the top of the gravity well between the Earth and Moon. Here transports from Earth and the Moon meet, swapping payloads etc. It is also an ideal transit point to Mars and the rest of the solar system. Come back in a hundred years, and this lonely outpost may have grown into a major port city in space. SKYCORP/Mark Maxwell

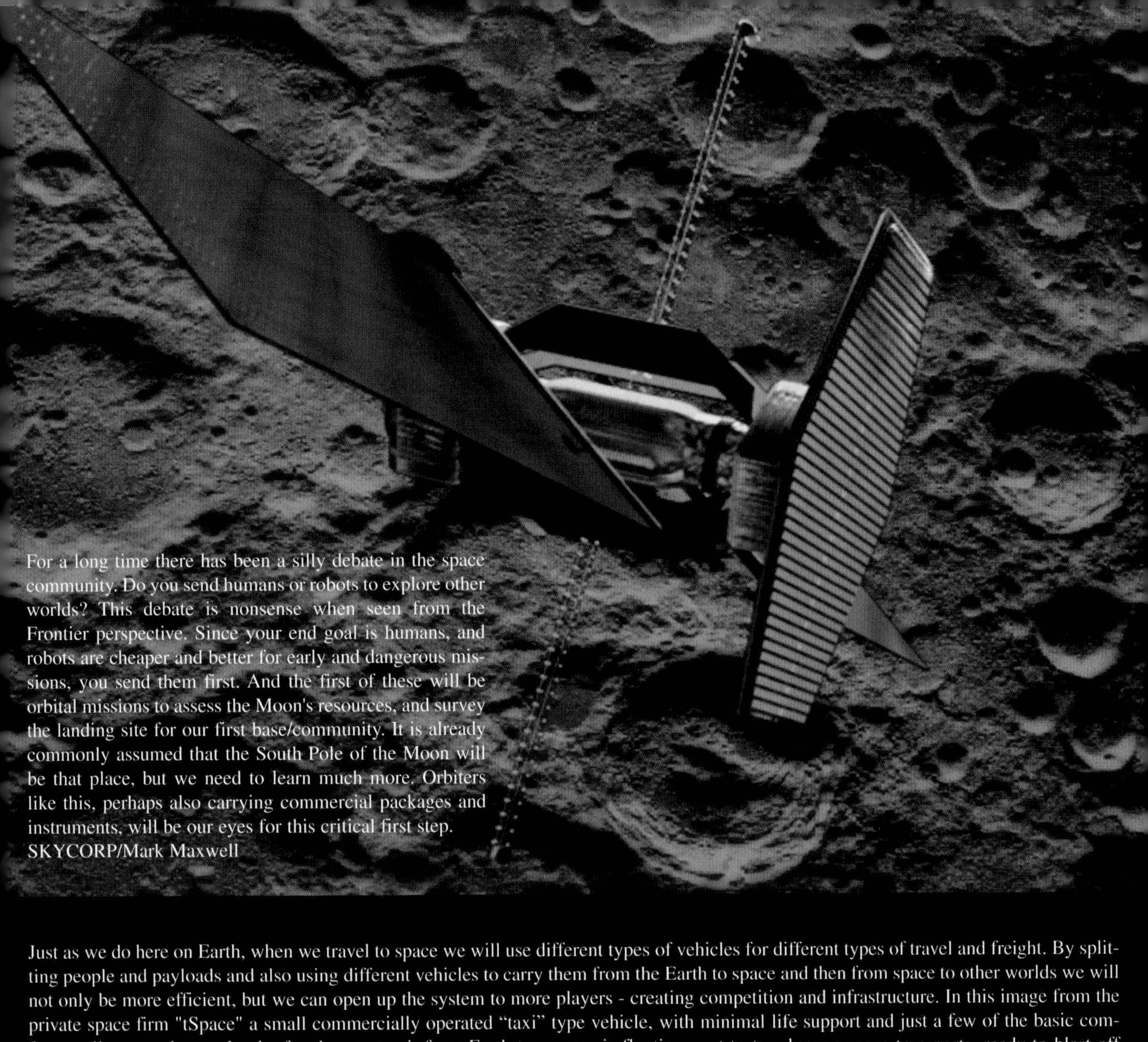

For a long time there has been a silly debate in the space community. Do you send humans or robots to explore other worlds? This debate is nonsense when seen from the Frontier perspective. Since your end goal is humans, and robots are cheaper and better for early and dangerous missions, you send them first. And the first of these will be orbital missions to assess the Moon's resources, and survey the landing site for our first base/community. It is already commonly assumed that the South Pole of the Moon will be that place, but we need to learn much more. Orbiters like this, perhaps also carrying commercial packages and instruments, will be our eyes for this critical first step. SKYCORP/Mark Maxwell

Just as we do here on Earth, when we travel to space we will use different types of vehicles for different types of travel and freight. By splitting people and payloads and also using different vehicles to carry them from the Earth to space and then from space to other worlds we will not only be more efficient, but we can open up the system to more players - creating competition and infrastructure. In this image from the private space firm "tSpace" a small commercially operated "taxi" type vehicle, with minimal life support and just a few of the basic comforts - all we need to make the few hours transit from Earth to space - is floating next to two human space transports, ready to blast off towards the Moon. t/Space-Mark Maxwell

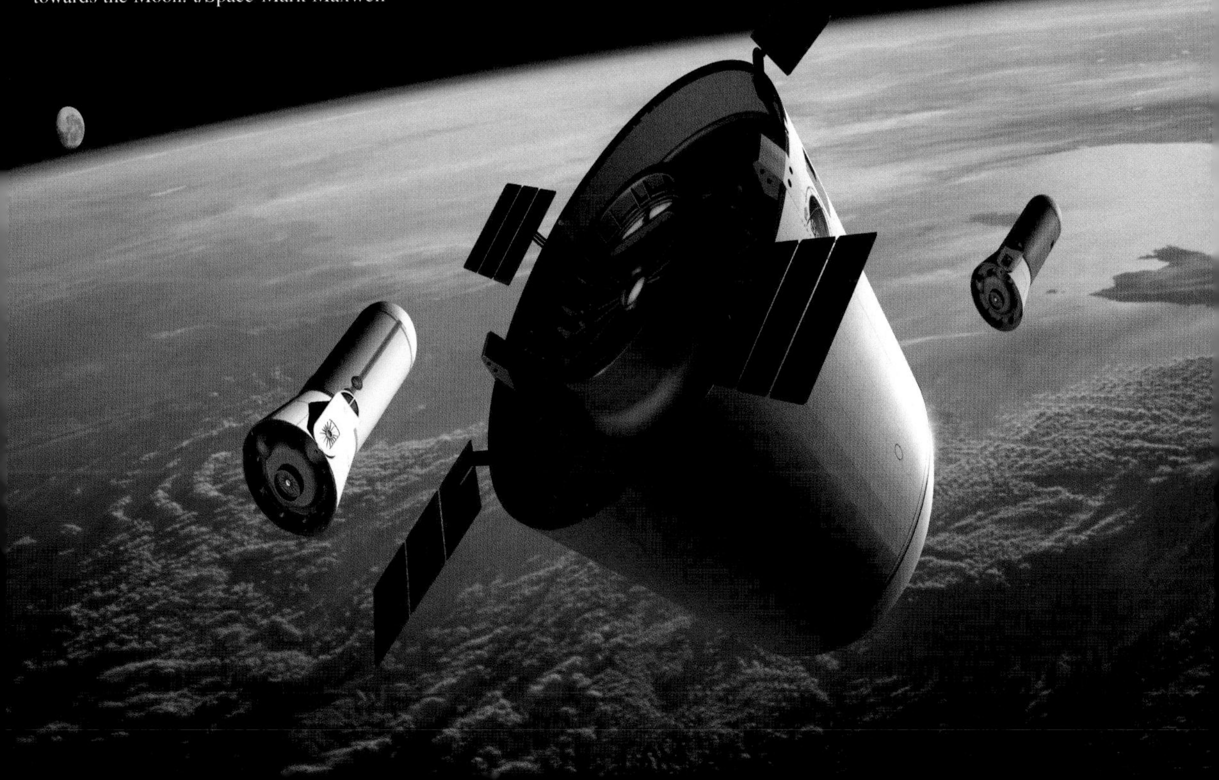

Switch (NASA/Pat Rawlings)
The LunaMod is now attached to the lifter and ready to become a space vehicle. Similarly shaped modules could be used for hardware and consumables, or brought on one-way trips to be assembled into habitat elements.

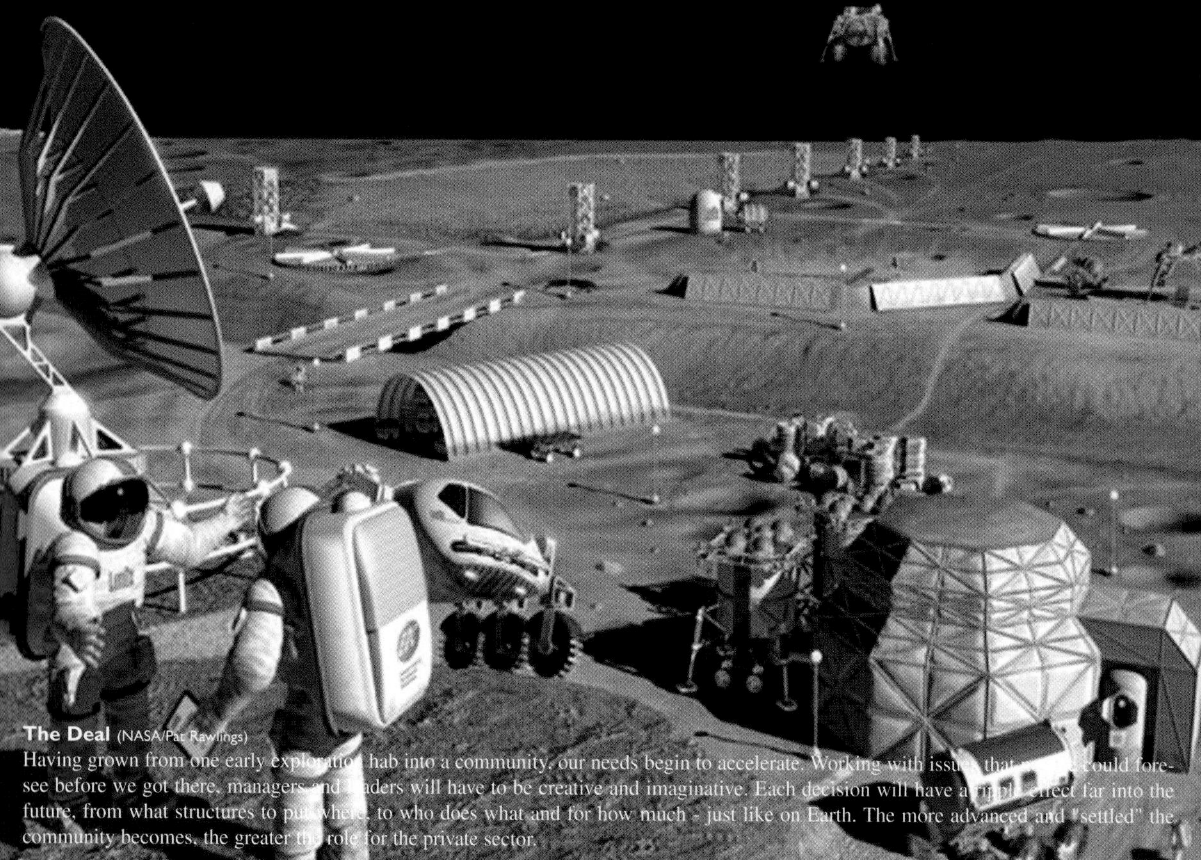

The Deal (NASA/Pat Rawlings)
Having grown from one early exploration hab into a community, our needs begin to accelerate. Working with issues that no one could foresee before we got there, managers and leaders will have to be creative and imaginative. Each decision will have a ripple effect far into the future, from what structures to put where, to who does what and for how much - just like on Earth. The more advanced and "settled" the community becomes, the greater the role for the private sector.

(NASA/Pat Rawlings)

Soon, the first outpost will begin to resemble a small community. From above habitat areas, landing pads and mining/industrial zones are now visible. As soon as possible, the use of in situ (local) resources will begin, and the newborn community will produce its own oxygen, water and propellant.

Apollo 11 Footprint (NASA)
This is one of the first human footprints on another world. Along with some hardware, flags and commemorative plaques, this is all that an alien visitor would find on the Moon to prove we had ever been there. Like the fossilized imprint of some long extinct dinosaur found on an ancient riverbed, it has become the symbol of failure, and is exactly what we do not want to have as the legacy of our next move outward from this planet.

there are several international agreements and treaties that have been effective in governing land and resources on and close to the Earth, which are not owned by any one country. Elements of these treaties can provide a framework for governing property rights in space.

OTHER INTERNATIONAL TREATIES & AGREEMENTS CAN SERVE AS MODELS

International Telecommunications Union

One example is the International Telecommunications Union ("ITU"), a specialized agency of the United Nations that administers both the geostationary orbital slots for satellites and the frequencies for satellite communications. The ITU-voting membership is comprised only of nation states; however, private entities can join the various sectors that correspond to their expertise. These entities are able to join in debates and the drafting of guidelines even though they cannot vote. The ITU serves as a trustee for the geostationary orbit, but, it has been criticized for using the first-come, first-served approach to handing out fixed satellite orbital slots, and because its structure allows those nations providing the least financial contributions to have the same level of voting power as those who contribute the most to the Union.

The ITU is a highly specialized agency, and provides numerous guidelines for nations and private sector members. Its' specialized sectors each have their-own advisory board and "study group" that examine potential issues. The ITU's legal framework is found in its constitution and the constitutions of its sectors, as well as the attendant regulations and the Optional Protocol. The Optional Protocol of the ITU specifies compulsory arbitration as a means of settling disputes among member states. The ITU is an example of international cooperation, but it does not face many of the challenges that must be overcome in the commercialization of space. For example, space commercialization will involve the removal of resources from celestial bodies as well as the construction of necessary infrastructure. However, the use of geostationary orbits, though a limited resource, is not as permanent, because the existing satellites can be removed and replaced with others. Additionally, the expense of placing a satellite into orbit is known, or at least capable of reasonable estimation, whereas space construction and mining of resources would be a novel undertaking. Thought to be very high, the actual cost of such operations is almost impossible to estimate at the present time.

The Antarctic Treaty System

A second example of an existing legal structure which could be transported to and applied in space is the system of management of Antarctica. The Antarctic Treaty System provides even more guidance for the governance of space resources than does the ITU. Like the Moon, Mars and asteroids, the Continent of Antarctica is also a vast expanse of land that is undeveloped and contains mineral deposits. The development and utilization of Antarctica, like the development of these celestial bodies, is expensive, requires great technical innovations and provides unique challenges to humans working in that environment. However, the development of and claims of sovereignty over Antarctica are restricted by a series of treaties known as the Antarctic Treaty System. Prior to the enactment of these treaties, several countries claimed portions of Antarctica. Those claims were then suspended by the Antarctic Treaty of 1959, in favor of a legal regime that protected the fragile environment and fostered scientific research in the region in the Antarctic Treaty of 1959. The Antarctic is governed by twenty-seven nations, known as "Consultative Parties," who gather annually and vote by a consensus on various matters. To become a Consultative Party, a nation must agree to the terms of the treaty and it must undertake "substantial research activity" on the continent. Other nations who do not meet the criteria may attend as "Observers" and participate in discussions; however, they may not vote on the issues.

The Antarctic Treaty System regulates scientific study, provides for the exchange of information between parties and provides guidelines for other management operations. Several provisions address the handling of waste and the protection of native species. The treaty also provides crucial guidelines for the safety and rescue of humans on Antarctica. It outlines a detailed plan for tourism, requiring advance notification and post-visit reporting as a means of monitoring human traffic. These provisions address areas of concern similar to those identified in Article XII of the Outer Space Treaty. The tourism resolution gives tourists a specific checklist of what they can and cannot do in Antarctica. Although these measures are in place, the Antarctic treaty lacks an administrative body to ensure the compliance of its members, but it provides for dispute resolution using negotiation or arbitration. Further, by agreement of all parties, any dispute may be brought before the International Court of Justice.

An international space governing body could perform similar functions. While the structure of the Antarctic Treaty is similar to that of the ITU and would seem to be an ideal model given its widespread acceptance and substantive provisions, it fails to deal with a crucial aspect of space development, the mining of minerals. In fact, a separate treaty, the Convention on the Regulation of Antarctic Mineral Resource Activities ("Antarctic Mineral Convention") was drafted to address this issue. However, this treaty has not been ratified by any nation. Instead, mining is governed by the Antarctic Treaty, whose stringent environmental protocol effectively prohibits any development of Antarctica's mineral resources by designating Antarctica as a natural reserve.

United Nations Convention on the Law of the Sea

There is an example of a legal regulatory regime that does address mining issues in a way that provides for commercial exploitation and that is the law governing the deep seabeds of the Earth's oceans. The seabed is rich in minerals that are found in secretions on the ocean floor as well as in the crusts of the deep sea. Collecting and mining these minerals is expensive and requires sophisticated technology capable of reaching the great depths. As in Antarctica, there are concerns for the environment.

The United Nations began drafting documents pertaining to the deep sea in the 1950s, but these documents did not deal with undersea mining because the necessary technology had not yet been developed. In 1982, the United Nations Convention on the Law of the Sea ("UNCLOS") was created to monitor the exploration of deep seabeds and oceans that are located farther than 200 miles from the coast of any nation. This area and its resources have been declared "the common heritage of mankind." UNCLOS created an International Seabed Authority ("ISA") to license and regulate the mining of this portion of the ocean. Unlike the Antarctic Treaty, membership in UNCLOS is not limited to those involved in active exploration, and each member may cast one vote. UNCLOS provides detailed regulations for deep sea mining. In addition to the regulations, UNCLOS created an intergovernmental mining company, Enterprise, to compete with the private entities granted licenses by the ISA. UNCLOS also allows for various means of dispute resolution, including adjudication by a specialized tribunal, the Seabed Disputes Chamber of the International Tribunal. However, because UNCLOS required mandatory transfers of technology, employed an economic model that preempted free-market enterprise, failed to assure access to future deep seabed resources, and included a voting structure that gave all nations equal control regardless of their technological capabilities or contributions to undersea exploration, the United States and other industrialized nations refused to ratify the 1982 agreement.

In an effort to add the industrialized nations to its membership, the UN renegotiated the mining provisions in 1994, creating the Agreement Relating to the Implementation of Part XI of the UNCLOS Convention. The United States took an active role in the negotiations, and as a result, the 1994 Agreement guarantees the United States a seat on the decision-making body, requires actual development by those mining companies granted a permit, and recognizes the current claims of those companies holding U.S. licenses. Transfers of technology are no longer mandatory. The United States signed the amended UNCLOS in 1994 and accepted provisional membership which was then extended in 1996. Despite these favorable changes, the United States failed to ratify UNCLOS and the incorporated 1994 agreement by the deadline in 1998 and lost its provisional membership. Though this amended treaty has had the support of presidential administrations from Ronald Reagan to George W. Bush, and has some Senate supporters, it has not yet been ratified by the United States Senate. In fact, until 2003, UNCLOS had not even been reviewed by Senators outside the membership of the Senate Foreign Relations Committee.

Deep Seabed Hard Mineral Resources Act

During the debate preceding the creation of UNCLOS in 1982, the United States passed its own Deep Seabed Hard Mineral Resources Act ("Seabed Act") to govern undersea mining. This Seabed Act makes clear that it is merely a temporary measure to be used until "a widely acceptable Law of the Sea Treaty is created, which will provide a new legal order for the oceans covering a broad range of ocean interests, including exploration for and commercial recovery of hard mineral resources of the deep seabed."

The Seabed Act states that the creation of an acceptable international regime, such as a modified UNCLOS, will not be accomplished very quickly. However, commercial mining operations would also take time to set up, and, therefore, technology development must begin as soon as they are able in order to begin mining when the minerals are needed. The Seabed Act proposes that the standard for exploration and commercial use of the seas should be a "duty of reasonable regard to the interests of other states," and that any uses should be in line with recognized principles of international law.

Similar to UNCLOS, the Act requires that undersea mining companies apply for permits and licenses to mine the deep seabed. It also describes instances under which such licenses may be revoked or renewed and provides environmental protection provisions, accident provisions and available legal actions. The Seabed Act also requires those companies in possession of a permit to diligently recover minerals within ten years of receiving their 20-year permit or the permit will be terminated. Such a provision ensures development of the area instead of dormant claims that leave an area unproductive. Though these provisions provide an extensive framework for mining, the Seabed Act advocates transitioning to an international mining regime.

In the Spring of 2004, Congress reconsidered the questionable provisions of UNCLOS, and the Senate Environment and Public Works Committee heard testimony advocating for the adoption of UNCLOS, which includes the 1994 Agreement. In the past, members of the Senate objected to the treaty because it lacked additional benefits not already enjoyed by the United States under the Seabed Act, while imposing additional obligations in terms of financial contributions to the ISA, and the possibility that a future decision made by the ISA would require the US to undertake additional commitments. However, at those hearings, it was argued that acceding to UNCLOS at this point would be fairly easy as it would not require the enactment of new implementing legislation, because the extensive Deep Seabed Mineral Resources Act already contains practices and regulations compatible with UNCLOS. The 1994 Amendment also gives the United States and other industrialized nations enough power to block or veto any new regulations that would further restrict mining. If the United States ratifies UNCLOS and the 1994 Agreement, the resulting international body could serve as a model for such a body to govern mining activities on the Moon, asteroids and Mars.

However, as the Deep Seabed Mineral Resource Act acknowledged, the creation and implementation of an international governing body can take many years, and in all likelihood would not be in place before the technology becomes a reality. Given the President's vision of sending another human expedition to the Moon by 2020, it seems likely that an international agency will not be created by that time. Therefore, other short-term measures must be created to bridge the gap.

INTERIM SOLUTIONS

An International Agreement Modeled on the International Space Station Intergovernmental Agreement.

For the short term, the International Space Station Intergovernmental Agreement ("IGA") provides the most workable model for a property rights regime in outer space. The IGA was signed in 1989 and has a "hub and spoke structure." NASA serves as the hub and has signed the agreement with the Canadian Space Agency, the European Space Agency and the Government of Japan. NASA then signed other bilateral agreements, called Memoranda of Understanding (MOUs) with other national space agencies to provide guidelines for the technical and administrative functions of the space station. In 1993, Russia joined the IGA and updated agreement was signed in 1998.

The IGA governs the use of the International Space Station ("ISS"), which is a research lab placed into low Earth orbit, for the purpose of conducting scientific research in a low gravity environment. The members of the IGA contribute funds and technology, and each owns some portion of the space station. The country with the ownership interest retains control of its particular physical module and its crew. The nation may contract with other countries that wish to use its portion for scientific research. The work that takes place on the module then remains subject to the laws of that nation and is considered to be within its jurisdiction. For instance, if scientific data is produced using an instrument supplied by the Russian space agency, then that data is subject to the patent laws of Russia. NASA serves as the coordinator for the various operations and provides the same oversight that would be supplied by an international organization, though it is not involved in dispute resolution. Disputes are adjudicated by either the International Court of Justice or the World Trade Organization, depending on the whether or not the particular claim involves international trade law.

Twenty years ago, NASA was asked to advance commercial activity in space. In 1998, Congress passed the Commercial Space Act, which directs NASA to use the ISS as a springboard for space commerce. The Act promotes the use of commercial launch services and emphasizes the importance of commercial providers in the operation, servicing and use of the space station. Though the Act provides some guidelines for space commercialization, it fails to adequately address liability issues, particularly insurance indemnification for commercial launches. Following adoption of the Act, NASA produced a "Commercial Development Plan" to implement its provisions. This plan calls for a non-governmental organization to manage future commercialization of space, but the plan description is almost silent as to how commercialization will actually be advanced by the organization. Much like the Deep Seabed Mineral Resources Act can serve as the implementing regulations for UNCLOS, the Commercial Space Act could provide the regulations for an expanded IGA.

The IGA could easily be applied to space tourism, settlement, development, and bases of operation on asteroids, the Moon and Mars. NASA could continue to serve as the coordinator unless a Non-Governmental Organization is agreed upon by the participating nations. Other countries would contribute funds and place technology on the Moon through their space agencies. These space agencies would secure the technology and funding from private businesses that enter into contracts for such services with these agencies. In accordance with the terms of the IGA and the Outer Space Treaty, each individual country, or space agency, would retain jurisdiction over its crew, its spacecraft and any structures or equipment.

Export Control Impediments

However, if the IGA model is to be employed as an interim legal system in space, one major impediment to the development of a property rights regime for the commercialization of space must be removed: the existing United States export controls on items that could have, potentially, both civil and military uses. To protect national security and U.S. foreign policy interests, the export of U.S. technology, commercial space products, services and commodities to the 15 partner countries involved with NASA in the ISS Program is controlled through a system of licenses issued by the Department of Commerce ("Commerce") or the Department of State ("State"). Under the authority of the Export Administration Act ("EAA"), Commerce established a licensing system to issue licenses for the export of dual-use items—those items that have both commercial and military applications—on the Commerce Control List. Similarly, under the authority of the Arms Export Control Act ("AECA"), and pursuant to the International Traffic in Arms Regulations ("ITAR"), State issues licenses and monitors and controls the export of military and dual-use technologies, goods and services. Commerce and State publish the Export Administration Regulations ("EAR") and ITAR, respectively, implementing the Acts. NASA must ensure that its ISS export activities conform to these laws and regulations, as it implements them. Given the overlap between the EAR and the ITAR, it is difficult to determine which controls are applicable in a given case.

The United States is also a party to a number of international agreements regarding dual-use technologies and goods. The series of agreements governing the ISS Program involving the 15 partner countries and their five cooperating space agencies provided the occasion to develop a comprehensive clause on the exchange of technical data and goods, which has since regularly served as a model for agreements covering other fields of space exploration. Certain ISS agreements are reduced to the simplified form of an exchange of letters.

Export controls restrict the ability of U. S. companies to freely share with, sell or convey to other nations commodities, technologies, goods and services relating to space. Under these laws, Items and services that could be used for military purposes are evaluated and then deemed to be defense articles or defense services if they do "not have predominant civil applications" or are not equal in form, fit or function to an established article or service used for civil applications. State and Commerce conduct case-by-case evaluations of articles and services which include a review of the nature, function, "variety, and predominance" of its civil applications as compared to the nature, function and possible capability of military use. If the article is found to be a "defense article," it is placed on the United States Munitions List, but only if "the failure to control such items … would jeopardize significant national security or foreign policy interests." These items include information systems and computer software that could become a "component" of a defense article. To sell an item on that list, the company must register with the Office of Defense Trade Controls and apply for a license. The arbitrary practice of classifying technologies and destinations results in an administrative practice of case-by-case evaluation of license applications with predictably arbitrary results.

The regulations state that launching a vehicle or payload is not considered exporting; but selling or transferring the contents of the vehicle or the vehicle itself "may" be subject to the controls. The regulations allow for temporary imports of dual use items to be brought into the United States from a foreign country and then later returned to that country. Critics argue that the regulations regarding export controls are too vague and reflect cold-war era alliances and enemies rather than today's global economy. For instance, the prize-winning SpaceShipOne may be grounded because, as a supersonic rocket, it is subject to the ITAR restrictions. These restrictions severely limit the ability of multi-national companies to develop and trade the sophisticated technology necessary for space exploration and development. The United States government is walking a tightrope between protecting the world and encouraging space development.

Additionally, if NASA continues to act as the coordinating agency and provide launch services, it may be very difficult to then transfer the launched materials to other countries and foreign companies in space. Continually seeking licensing for what could become routine work would be cumbersome, expensive and ultimately inhibit the development of space. Alternatively, companies would look to the launch services provided by other space agencies, such as the Russian, European or Canadian Space Agencies, as a means of launching dual-use technologies into outer space. In that event, United States companies providing the same services would lose their market share.

If the ISS/IGA model is applied to celestial bodies in space, with individual countries retaining jurisdiction over their crew and space objects, items brought from a foreign country to the United States for launching would return to the country from which they came if they are sent to an area in space under the control of that country through the IGA or its MOUs. It is possible, therefore, that these items could be considered merely temporary imports. However, an import license would still be needed, unless the items fall within an exception. Further examination of this idea is necessary given the complexity of the ITAR regulations and the wide discretion granted to State in classifying articles.

Exclusive Economic Zones

While the IGA could serve as a springboard for a property rights regime in space, there is another property management system already in use on Earth that incorporates many of the same principles of the IGA, but is more expansive and sensitive to the needs of individual nations. Despite failing to ratify UNCLOS, the United States opted to create an Exclusive Economic Zone, a concept found in Part V of UNCLOS. Under UNCLOS, a country may declare an area between its coast and 200 nautical miles as its "exclusive economic zone" (EEZ). The country then has the exclusive right to explore, exploit, conserve and manage the natural resources found in this area, including the resources found in the seabed. The country may also construct artificial islands and other installations or structures in its EEZ as long as doing so does not interfere with established sea lanes or otherwise compromise the safety of other ships using the waters. However, other countries must be allowed to navigate through the waters, fly over the area and lay pipelines or other cables on the seafloor in accordance with other international treaties. Ships passing through the EEZ of another country may not conduct research, catch fish, pollute the area or in any way take resources from that EEZ except in the case of an emergency. A country may also grant licenses or permits to other nations to fish the waters or make other uses of the resources found there or impose quotas or taxes to limit foreign fishing. UNCLOS encouraged countries to form regional and bilateral agreements with landlocked nations and other countries who cannot claim EEZs of their own.

Currently, the United States EEZ has not been open to offshore commercial "aquaculture" such as the development of extensive fish farming techniques or "ocean ranches" in the open waters. There are several reasons for this, including a lack of clear regulations and permitting procedures, existing legislation restricting foreign investment and use of the United States EEZ and the expense of developing extensive structures in the deep sea. However, both private and public research has been ongoing in this area, including Sea Grants funded by the U.S. Department of Commerce through its National Oceanic and Atmospheric Administration (NOAA) to study the offshore aqua culture of fish and other marine life for commercial use.

In addition to funding experimental projects, NOAA is drafting "offshore aquaculture legislation" that will address many of the impediments to commercial aquaculture. This legislation will establish a permitting process, create long-term leases for such activity, and provide exemptions from existing legislation so that foreign companies may obtain leases and invest in aquaculture in the United States EEZ. This legislation was expected to be sent to Congress in 2004, but as of December 2004, it had not yet been filed. As an interim measure, NOAA has created a voluntary Code of Conduct for commercial aquaculture, which provides guidelines for companies. However, a final draft of the Code has not yet been published. There is support in the Senate for strengthening and redefining US policy regarding aquaculture. In June 2003, a resolution was introduced in the Senate calling on the federal government to promote aquaculture and to achieve five times as much aquaculture production by the year 2025.

EEZs could be created in space, giving each nation the option of building a structure on a celestial body or occupying an orbit with spacecraft, and then claiming up to a certain amount of area around their structure or craft for their use. As mentioned above, countries participating in the ISS have already secured small safety zones around their vessels. The amount of area to be claimed would have to be agreed upon prior to occupation. These zones would be modeled after those on Earth, allowing the nation to contract or lease portions of the area to other nations, such as those countries, which lack the technology to launch spacecraft into outer space. The EEZ of one nation would allow other nations to pass through to its own EEZs so long as those countries did not disturb or remove resources as they move through the area.

Additionally, each nation would retain jurisdiction over its EEZ and could create its own regulations and permitting procedures. If the ITAR provision discussed above allows for temporary imports to be transported back to the country of origin without special licensing requirements, then items from other countries could be shipped from the United States to portions of the US EEZ in space that have been given to those nations through long-term leases or licenses.

While NOAA retains control over the use and licensing of the US EEZs on earth, an international government organization, similar to INTELSAT, ITU or UNCLOS may be a more appropriate regulatory body to manage the utilization of property and economic development in space. It has been possible to devise in the ITU an international organization capable of dealing with allocation and sharing of electronic spectra. It should be no more difficult to develop a pragmatic infrastructure to stimulate and expand the space economy. No single model necessarily fits the needs of a future special or general international space organization. There would be a need to accommodate the views of nations with space resources and those in the process of development.

CONCLUSION

Many countries with government space programs are rapidly becoming technologically and economically capable of implementing a viable space industry. Companies and entrepreneurs play an integral role in this multi-billion dollar enterprise. However, a comprehensive legal system governing operations on celestial bodies does not yet exist. Substantial investments of capital are needed to launch a viable space industry on the Moon, Mars and asteroids. Stimulation of the massive investments required for such commerce can only occur where there is a reliable system governing legal rights and obligations.

In the future, it may be more desirable to create an international governing body to administer a legal system governing celestial bodies, by combining and refining elements of the international regimes described above, particularly that of the ITU, IGA and UNCLOS if it gathers enough support from the industrialized nations.

Such a body would handle the issuing of licenses or permits as well as develop guidelines for space exploration, mining of resources, accident liability and legal claims to be resolved, perhaps by a specialized Space Tribunal. Such a governing body and legal system must also avoid the pitfalls of the failed attempts to successfully develop Antarctica and the deep seabed. Since the President's year 2020 deadline is rapidly approaching, the foundations of a legal system must today make that giant leap into outer space.

Rosanna Sattler is a partner with the Boston law firm of Posternak Blankstein & Lund, LLP, and chair of the firm's Space Law and Telecommunication Group, as well as its Environmental Group, with extensive knowledge about a range of emerging legal issues in the commercialization of outer space. She was a presenter at the International Space Business Council Law Forum at the National Press Club. In 2003, she authored an article in Kluwer Law International's Air & Space Law concerning U.S. commercial activities aboard the International Space Station. Ms. Sattler's interests include the developing law of contract and property rights in space, as well as liability and insurance for space endeavors. A Harvard Law School graduate, Ms. Sattler is a member of the American Bar Association's Forum on Air and Space Law.

Alan Wasser was one of the first people I met when I began to be active in the space settlement community. For years he been a stalwart proponent of human communities on the Moon, and was the first person to propose that such a settlement might be located on or near a permanently lit mountaintop on the Moon's south pole. (He was given credit for this in Ben Bova's book "Moonrise" in which the mountain is named Mount Wasser.) Alan has also been pushing the idea of lunar property rights for over a decade. He and many others believe that if handled with care, the key to opening space to humanity will be the right of individuals to own their own land—something we take for granted in free societies here on Earth, but which, for reasons he outlines in the following pages is almost seen as heretical by some when it comes to space. To many in the space movement, including myself, the real insanity is to not understand just how small a speck of land Earth itself represents when compared to the rest of the universe, and based on this near sighted hubris to proclaim we should not go to and live on other "lands" in space—as if this were all there is and ever shall be. This is especially true since places like the Moon are dead rock, not being taken from anyone, and will be given to future generations as their new homes.

The Space Settlement Initiative

by Alan Wasser

Will people ever live and work on the Moon and Mars? Will the settlement of space take place in your lifetime?

The settlement of space would benefit all humanity by opening a new frontier, to energize our society, provide room and resources for the growth of the human race without despoiling the Earth and create a lifeboat for humanity that could survive even a planet-wide catastrophe. Unfortunately, as things stand now, space settlement will not happen soon enough. I believe the "how to" get space settled lies in creating a sufficient "why to" for private enterprise.

Space development has almost stopped, primarily because no one has a sufficient reason to spend the billions of dollars needed to develop safe, reliable, affordable transportation between the Earth and the Moon. Neither Congress nor the taxpayers want the government stuck with that expense. Private venture capital will support such expensive and risky research and development, only if success means a multi-billion dollar profit. Today, there is no profit potential in developing space transport, but we have the power to change that.

There is a "pot of gold" waiting on the Moon, to attract and reward whatever companies can be the first to assemble, and risk enough capital and talent, to establish a "space line" and Lunar settlement. Legislation proposed by the Space Settlement Initiative would give private enterprise a way to make the immense profits necessary to justify immense effort and expense to its investors. It would save NASA and the taxpayers the cost of developing affordable space transport, by allowing private enterprise to assume the burden of settling space. In addition, ordinary people could purchase tickets and visit the Moon as tourists, scientists, or entrepreneurs. We could create vast wealth by making it possible for a settlement to claim and own—and re-sell to those back home on Earth—the product that has always rewarded those who paid for human expansion: land ownership.

Currently, Lunar and Martian real estate is worthless. But that real estate will acquire enormous value after there is human settlement, regular commercial access, and a system of space property rights. Lunar or Martian property ownership could then be bought and sold back on Earth, raising billions of dollars. This is a plan to be sure that money is used as an incentive and reward for those who invest in a way to get there and stay there.

In the mid 1960's, Pres. Lyndon Johnson saw he was going to be forced to take money from the space race to fund the Vietnam War. He feared if that allowed the Russians to win the race to the Moon, they might claim ownership of it. So he proposed, and the US Senate ratified, what became known as the 1967 "Outer Space Treaty." Among other things, this treaty prohibits any claims of national sovereignty on the Moon or Mars, etc. Therefore no nation can claim or "grant" land in outer space.

But, quite deliberately, the treaty says nothing against private property. Therefore, without claiming sovereignty, the US could recognize land claims made by private companies, regardless of nationality, that establish human settlements on the Moon or Mars. The US wouldn't be "granting" or giving the land to anyone. It isn't ours to give. The settlement says "because we are the first to actually occupy this unowned land, WE claim ownership of it"—and the US "recognizes"—accepts, acquiesces to, decides not to contest—the settlement's claim of private ownership.

The dollar value of a Lunar land claim will only become big enough to be profitable when people can actually get to the land. So Lunar land deeds, recognized by the US under this plan, can be offered for sale only after there is a transport system going back and forth often enough to support a settlement, and the land becomes accessible. It will finally be understood to be land in the sky, not pie in the sky.

The proposed legislation would commit the US to granting that recognition if those who have established settlements meet specified conditions, such as offering to sell passage on their ships to anyone willing to pay a fair price. Entrepreneurs could use that promise of US recognition to help raise the venture capital to develop the ships needed to make the claim.

It would take a really large claim to be worth that huge investment, of course, but there is an amazingly large amount of land out there waiting to be claimed. For example, a claim of 600,000 square miles, about the size of Alaska (just under 1,600,000 square km.) would be only around 4% of the Moon's surface, but would be worth about $40 billion at even a very conservative price of $100 an acre (4047 square meters). The price of the land might, by then, be much greater.

It could be offered for sale after months of worldwide press coverage produced by the race to be the first to settle the Moon. There will be land buyers with business purposes for buying and using land, but there will be a much bigger speculative and investment market. Many people who will never leave Earth will buy Lunar land—some in hopes of making a profit, others just to be part of the excitement or to leave an acre to their grandchildren. Some may even put their name on a crater.

The profits on land sales which take place in the US will, of course, be subject to US taxes, so the Budget Office will score this legislation as a revenue producer, not a cost to the US. It sounds strange because we haven't done it yet, but there is growing sentiment for extending private property and the benefits of free enterprise to space. Former House Science Committee chair Bob Walker has suggested that the Bush administration would like to develop such a legal structure.

Before copyright and patent laws, no one could own songs, stories or ideas. The passage of those laws, creating intellectual property, made whole industries possible and added greatly to the world's wealth from things that had previously been valueless. Creating Lunar property could be the incentive to open the space frontier to everyone, benefiting all of humanity.

Land Claims Recognition Law

The creation of a legal system of property rights for space is not the long-term objective. The establishment of a property rights regime for space is only a means to an end, not an end in itself. The real purpose is to enable the expansion of the habitat of the human species beyond the Earth by offering a huge financial reward for privately funded settlement. It is the only way to create an economic incentive sufficient to encourage private investment to develop affordable human transport to the Moon and Mars.

There are alternative space property rights schemes being proposed by some lawyers that would, instead, make settlement even harder than it would be now. They would require that, if you do pay to develop space transport, you would then have to pay the UN or some other body even more for the land you want to settle.

Property rights legislation should be judged by how well it encourages space settlement, not on how elegant the resulting property rights system is. Property laws could be left to evolve after settlement, except that settlement just isn't happening without them, so we need something like this legislation to jumpstart it.

Affect on Space Transport

Hopefully, promising property rights will turn out to be enough to produce the necessary investment. But it is impossible to know, this early, whether it will be. After all, it is impossible to know now how much it will cost to develop safe, reliable, affordable space transport, or how long it will take.

There is also no way to be sure just how much Lunar land will be worth when recognized deeds are being sold by people who can actually take you (or your customers) to that land. But a piece of Lunar land the size of Alaska would certainly be worth a very large amount of money.

Those who say property rights are not needed until after settlement has taken place, are counting on near-term incentives (such as space tourism, servicing the space station, etc.) to produce all the necessary investment in affordable space transport, the establishment of on-orbit infrastructure and then settlement itself. It is very much open to question whether such near-term incentives could be sufficient, but it is certain that adding a very big long-term incentive, on top of whatever near-term incentives there are, would have to help.

Imagine that you are an entrepreneur trying to get a venture capitalist to fund your research on a radical new idea that you think might reduce launch costs by an order of magnitude or more. He asks, "If you succeed in this risky venture, how are you going to use it to make enough profit to make it worth my while?" You tell him your projections of space tourism profits, etc., and he is impressed, but not enough. Then you add: "In addition to all that, if we do reduce launch costs enough, it could later be used to establish a settlement on the Moon and immediately gain US recognized ownership of 600,000 square miles that could be sold, and/or mortgaged, starting the very next day. If that were valued at only $100 an acre, it would be an instant gain of a $40 billion dollar asset on your books." Perhaps this might help your case a bit.

In sum, in order to spur the development of affordable space transport, this law doesn't need to bring in all the needed investment by itself. There are existing incentives, but not quite enough. We need only bring in sufficient additional financing to tip the balance. The promise of property rights for space settlement is a very low cost, low risk, "do-able" way to attract that supplementary venture capital.

International Law

Early in the negotiations for the 1967 "Treaty On Principles Governing The Activities Of States In the Exploration And Use Of Outer Space, Including The Moon And Other Celestial Bodies", generally called the Outer Space Treaty, the USSR suggested that the treaty ban private activities in space but, at the insistence of the Americans, all such provisions were dropped from the final treaty. According to the New York Times report of the US Senate ratification hearings for the Treaty, (March 7, 1967) Senator Albert Gore (Senior) worried that the "benefit of all" provisions of Article 1 of the treaty might inhibit space activities. The Times says Arthur Goldberg, who negotiated the treaty for the US, reassured Gore by describing "the article as a 'broad general declaration of purposes' that would have no specific impact until its intent was detailed in subsequent, detailed agreements."

The one serious attempt to establish such a follow-up agreement was a disaster that the US Senate refused to ratify, specifically because it attempted to ban private property. It was the 1979 "Agreement On The Activities Of States On The Moon And Other Celestial Bodies," generally referred to as "The Moon Treaty." It would have replaced the "benefit of all mankind" language with the drastically different "common heritage of mankind" doctrine. Some third world countries have claimed that the "common heritage" doctrine would mean that anyone wanting to establish a Lunar settlement might have to pay off the leaders of every nation on Earth.

Fortunately, since it wasn't ratified by the US or any other space-faring nation, the Moon Treaty is generally regarded as a dead letter, and is not binding on the US or its citizens. As things stand now, private entities can claim ownership of land on the Moon "on the basis of use and occupation," although nations cannot. Resistance to this idea in the State Department comes from those who sincerely believe only governments and government employees belong in space at all, ever, as a matter of principle and safety.

National Sovereignity

In France, which follows "civil law" (as opposed to "common law" which the US inherited from the U.K.), property rights have never been based on territorial sovereignty. Instead, they are based on the "Natural Law" theory that individuals mix their labor with the soil and create property rights independent of government. Government merely recognizes those rights.

Throughout history, actual settlement (defined as occupation and use) has been the traditional basis for claims of ownership of land that had no sovereign. Christopher Columbus claimed the land he discovered by leaving a garrison on it, not by planting a flag. We want the US to treat the settlement, itself, as having one of the attributes of a sovereign: the right to claim private ownership of unowned land by right of use and occupation.

For property rights on the Moon, the US will have to recognize Natural Law's "use and occupation" standard, rather than the common law standard of "gift of the sovereign", because the common law standard cannot be applied on a Moon where sovereignty itself is barred by international treaty. The US will have to say that, because there can be no government on the Moon, a true settlement can give itself title, just as though it were a government, and its property deeds, for land under its control, will be recognized by US courts of law, (subject to specified limitations) just as titles issued by France, China and even Iraq, are recognized by US courts.

Because the US market represents such a large fraction of the world's economy, and because it often leads the way on economic matters, US recognition is by far the most important—and the place to start. But it certainly would be very desirable if other nations joined in, especially those with significant space industries, such as the members of the European Union, Russia, Japan and China. Therefore, it is important that those nations see more benefit to themselves in joining than resisting.

The legislation I propose strongly encourages reciprocal arrangements with other nations. It instructs the State Department to actively seek those agreements. If needed, it allows State to negotiate treaties that require that settlements be multi-national consortia, to assure other nations that this isn't going to be just an American land grab. If necessary to get the UN on board, it even allows State to negotiate treaties requiring the inclusion of citizens of at least one developing country as investors or providers of an equatorial launch site.

Will this be enough to guarantee all nations sign on? Probably not at first, but it won't really make a significant difference to land buyers if North Korea and Cuba, etc., refuse to recognize their land deed, as long as they know the US and the major spacefaring nations will.

It would be nice if we could offer a series of graduated rewards for each little advance in space development, but it can't be done legally, and the grants wouldn't be worth anything if it could be done. Where the US has sovereignty, and is the source of ownership, the government can give ownership of land, or limited rights to its use, for whatever reasons it chooses. But, since no nation can claim sovereignty on the Moon and Mars, the US has nothing to give. The only thing governments can do is to recognize, or not recognize, a claim made by a private entity which has a good case for making the claim.

This law would not prohibit anyone from making a claim to any space real estate based on anything, or nothing at all, including "I want it, so it is mine". Nor would it require anyone else to pay any attention to such a claim. It would only require that the US government must recognize a claim based on actual settlement and "use and occupation".

It will take hard work to get Congress and the courts to accept even settlement and "use and occupation" as a basis for space land claim recognition, even though that has always been the basis for claims of ownership of new land. Space claims based on anything less than settlement would be virtually impossible to justify to the courts and the world.

More important, human settlement of space is our real goal! We are a lot more likely to actually see it happen if it is the required condition to win anything. Giving limited ownership for less could reduce the incentive, for both the winners and losers of the first round, to keep going full out toward settlement. Only when there is a live human being waiting on the Moon for the return flight can we be really sure that there will be a return flight, even if the accountants say, "put it off for a few years, or more."

Lunar Worth

It isn't how much land you get that matters, it's how much it's worth! The dollar value of an acre of Lunar land goes up exponentially the day buyers can actually buy a ticket and go there, or send a representative or a customer. That means the value of the grant goes up exponentially if we hold it back until there is a space line going back and forth.

The value of the land claim can be similarly increased if we capitalize on the media coverage of a space ship taking off to try to win the race to establish the first human settlement on the Moon. The day people land on the Moon, set up permanent habitation, and stay there while the ship goes back for more people, they will be the whole world's heroes. At that very moment, back on Earth, their representatives will finally be free to start selling seats on subsequent trips, at prices people all over the Earth can afford and valid deeds for land around their base. People will even buy land as a way to support, or just feel part of, the project.

Then, and only then, will a Lunar land claim reach the multi-billion dollar value that would make a real difference, enough to justify the billions it took to win it.

We should set an appropriate limit to the amount of land that can be claimed, (and it will be easier to increase the size of a grant, later, than to reduce it). We should require the settlement to behave by international norms. We should require that the settlement be open to all and prohibit anti-competitive behavior. Regulations could even include protection for sites of historical or other special importance.

It might also be required that only a certain percentage of land sale revenue can be used to repay the cost of establishing the settlement and taken as profit, the balance being retained to support the settlement itself until it can find ways to earn enough to become self sufficient.

Property Claim

The first settlement on the Moon should be able to claim up to 600,000 square miles. Getting to Mars will cost much more and Mars itself is larger than the Moon. Therefore the first Martian settlement should be able to claim up to 3,600,000 square miles, roughly the size of the United States, worth 230 billion dollars at even $100 per acre.

Some critics object that would allow a settlement to claim more land than it can use, but the amount of land that can be used depends on what you are using it for. Nineteenth century land grant farmers used 40 acres and a mule. Modern mechanized farms use vastly more land than that. Cattle ranchers use much more land than farmers. But none of those are the size criteria that should be used for a Lunar settlement because, of course, the settlement will not make its living by either farming or ranching.

The plan is to let settlements recoup the cost of getting there in the first place by selling land. If you are in the real estate business, especially if you are selling totally raw Lunar land, you can use all the acreage you can get title to. So the "right" size for a claim is that size which is large enough to justify the cost of developing reliable space transport and establishing a settlement, and small enough to force the development of cost effective, affordable, transport, and small enough to still leave room for future settlements.

That's how the proposed settlement sizes are derived. Real estate experts have estimated the minimum the land would bring when you can buy a ticket and get to it. Space experts estimated what was the least that financially efficient private companies could hope to establish settlements for. The average settlement cost estimates, divided by the estimated average dollars per acre, gives the number of acres needed. Converted to square miles, that works out to approximately 600,000 square miles on the Moon and 3,600,000 square miles on Mars.

Fortunately, that is quite small enough to still leave plenty of room for subsequent settlements, since it is only around 4% of the Moon, 6.5% of Mars. Since it is much easier to follow than to lead, and we want to encourage leadership, no settlement after the first gets even that much. Each subsequent settlement gets 15% less land than the previous one. Finally, no entity can get recognition for more than one settlement on a body. Therefore there is no possibility of anyone monopolizing all the land.

International law clearly requires that opening the space frontier must "benefit all mankind" and that there be "access to all areas of celestial bodies." The Outer Space Treaty in its very first article, says, "the exploration and use of outer space, including the Moon and other celestial bodies, shall be carried out for the benefit and in the interests of all countries ... and shall be the province of all mankind. ...and there shall be free access to all areas of celestial bodies."

Article XII says: "All stations, installations, equipment and space vehicles on the Moon and other celestial bodies shall be open to representatives of other States Parties to the Treaty on a basis of reciprocity. Such representatives shall give reasonable advance notice of a projected visit, in order that appropriate consultations may be held and that maximum precautions may be taken to assure safety and to avoid interference with normal operations in the facility to be visited." All settlements and property owners will have to accept that rule unless the Treaty is ever changed.

Establishing a space line and settlement open to all paying passengers, regardless of nationality, would certainly benefit all mankind, making it both necessary and sufficient to meet that very important condition of international law.

The question of compliance with the access requirement of the Outer Space Treaty upsets a lot of people, on both sides of the issue. Some call me a Communist for acceding to the access requirement, it means property owners can't have an absolute right to keep everyone else off their land. Others insist my plan fails because it does not go far enough to comply with that same "free access" rule. I, of course, think I've made the best compromise possible until and unless the treaty is revised.

The United States will probably be the first market where land deeds will be sold to the public, and it will be the US courts that will rule on whether Lunar land sales are valid transactions or frauds. What this legislation does is tell the US courts what standard to use in making that ruling. Further, it is not at all unusual for quite a few other nations to follow the US's lead on things like this. However, this legislation is most definitely not just for the benefit of Americans!

Given today's global economy, it is almost certain that all entrants in the race to establish a settlement will be multi-national consortia. The investor/owners will be drawn from all around the world, as will the land buyers. Most particularly, the teams of aerospace companies cooperating to build the ships will be from many nations. It is too big a job for one company, or even one nationality, to undertake alone.

This is much more than the work of only one person. Many others have contributed a great deal to it, deserving thanks and credit for supplying key ideas, explaining the fine points of international law, promoting the idea, helping with writing and editing or in other ways, and even by attacking the plan and exposing weak points that needed fixing. Among the many who contributed are Scott Pace, Marianne Dyson, Glenn Reynolds, Douglas Jobes, David Wasser, Declan O'Donnell, Ray Collins, Arjen Van Ballegoyen, Leonard David, Lawrence Roberts, Ben Bova, Rick Tumlinson, Carol Kochman, Toni Sonet, Art Dula, Robert Zubrin, Jim Bennett, Bob Werb, Jeff Krukin, Charles Wood, Fred Ordway, Jim Benson, Ed Wright, Alford Lessner and Charles Miller. Sincere apologies to the many others who should have been mentioned but have been accidentally overlooked.

Alan Wasser was a former broadcast journalist at ABC News and CBS News, who then owned and operated a successful international business, which he sold. He is the Chairman of the Board of the Space Settlement Institute, http://www.Space-Settlement-Institute.org/ . He was the Chairman of the Executive Committee (CEO) of the National Space Society and is now a member of its Board of Directors. He was a member of the Board of Directors of ProSpace, and is an Advocate of the Space Frontier Foundation. He was the first to propose that the first human settlement on the Moon might use a permanently sun-lit mountain top at the moon's south pole, the existence of which was only later confirmed by the Clementine mission. Much of Ben Bova's novel "Moonrise" takes place on that lunar mountain, which Bova named "Mt. Wasser". He is the originator of the idea of using land claim recognition to make privately funded space settlements potentially profitable, and therefore possible in our lifetime. He is the author of numerous articles on the subject of space property rights, most recently in The Explorers Journal, the official magazine of the Explorers Club, Space News, Ad Astra, Space Governance, Space Times and Space Front among others. Space Governance published a rough draft of the proposed legislation.

Tradition is the enemy of innovation. As young as our space program is, we have already evolved many traditions that must be overturned if we are to open the frontier. For example, the idea that space probes must be huge projects, with large staffs, lots of managers and government oversight etc. Since the first steps of exploration leading to human habitation are going to be done robotically it is important that these missions return as much information as possible for the dollar spent. (By the way, there is no conflict here as some would have you believe—robots/probes go first, then people follow!) Lean, mean explorations teams are the answer. Whether working on government projects, or responding to government offers to buy their data, such teams allow for more versatility, robustness and redundancy than giant all in one missions. As an employee of the Space Studies Institute (where the idea was conceived) I was lucky to have been one of the first people to join Dr. O'Neill's push to look for water on the poles of the Moon. Having just finished proving that an electromagnetic mass driver could reach high enough velocities to hurl payloads off the Lunar surface, SSI wanted a new showcase "critical path" project. Prospector was chosen and off we went. It is important to note, as Dr. Binder mentions, Prospector was the first ever citizen initiated space probe. We literally raised the money ourselves to get the initiative going, doing roadshow presentations at space conferences, and people contributing ten and twenty dollar donations. Dozens of grass roots activists across the country worked countless hours to push the idea. A recently deceased musician friend of mine, Tim McGlashin, even put together a "We are the World" benefit concert in Houston at the time, to help us keep the idea alive. These volunteers kept the Lunar Prospector torch lit for several years, until our white knight arrived, in the form of Dr. Alan Binder. After several years of private effort, Al finally turned to a new funding initiative put out by NASA, and took the project to a whole new level, championed it and made it happen. The rest, as they say, is history.

Lunar Prospector: Lessons Learned and the Way to the Moon Commercially

by Alan Binder, PhD

When the idea of doing a private, lunar orbit mapping mission to the Moon reached its critical point in November of 1988, those of us who were attempting to do the impossible had several goals in mind. We wanted to A) show that inexpensive, simple missions with limited objectives could be done outside of the huge and inefficient NASA bureaucracy and produce high quality data, B) demonstrate the feasibility of commercial lunar exploration, C) provide data on lunar resources that could benefit humanity on Earth, and D) rekindle interest in the exploration and colonization of the Moon.

First it is important to note that Lunar Prospector was developed as a private effort, completely outside of the NASA hierarchy and hence completely unencumbered by it. Six years before I proposed Lunar Prospector as a Discovery Program Mission, the Space Studies Institute in Princeton, NJ, several Chapters of the National Space Society and the Space Frontier Foundation, along with my colleagues and I, developed Lunar Prospector as a private effort to demonstrate that small, inexpensive spacecraft with limited payloads could make major contributions to the exploration of the Moon and planets.

After several years of effort to find private support for the mission, I asked my employer, Lockheed, if it would support my proposal to the new NASA Discovery Program, in which the missions were to be conducted by the Principal Investigator and whose goal was to find ways of doing missions "Faster, Better, Cheaper". Lockheed said yes and in February 1995 Lunar Prospector was chosen as the 1st peer reviewed and competitively selected Discovery Mission, the only one ever chosen directly for construction and flight and the least expensive ($63 million) by a factor of four.

Between January 6, 1998 and July 31, 1999, in my role as Mission Director, I flew a completely successful 19-month polar orbiting mapping mission that produced data, which far exceeded that promised NASA at the beginning of the program. However, the real success of Lunar Prospector was that I was able to prove all the programmatic goals we had established in the beginning, thereby showing how to get to the Moon commercially—without the interference of NASA. For those of you who are interested in the full story of the program, it is documented in my book, *Lunar Prospector: Against All Odds*, published by Ken Press (520-743-3200, also see my Institute's web site, Lunar-Research-Institute.org). The "Lessons Learned" from Lunar Prospector are summarized here in my report to the FINDS Endowment committee.

Following the success of Lunar Prospector in demonstrating the commercial feasibility of doing commercial lunar exploration missions and the passage of the "*Commercial Space Act of 1998*", I have been lobbying for a NASA administrated Lunar Data Purchase Program that would allow commercial missions to be flown. My rational for such a data purchase program is given in the second section, followed by a short review of lunar resources and how they might be used.

LESSONS LEARNED - COMMERCIAL LUNAR MISSION COST ESTIMATES

The following reports on the estimated costs of Commercial Lunar Missions is based on the author's experience as the Principal Investigator of the highly successful Lunar Prospector (LP) Polar Orbiter Mapping Mission. It must be noted that LP was conceived as a private mission in the late 80's as a demonstration of the viability of small, inexpensive missions for doing lunar science and as a demonstration of the viability of such missions as commercial ventures. Because LP was conceived as a privately supported mission, every effort was made to keep the spacecraft and mission as simple, as reliable and as inexpensive as possible—goals that any commercial company must also have if it is to be financially successful. Thus, unlike any other lunar or planetary mission flown to data, LP serves as a near perfect model for both commercial lunar exploration ventures and for estimating the costs of such missions. However, since LP was conceived as the simplest and least expensive lunar mission that would produce (and did produce) valuable science data, its cost represent those of the least expensive of the commercial missions—all other will cost more than LP.

Though LP was intended to be a privately supported and privately conducted mission, the several years effort to obtain the financial support for the mission failed and the author then proposed the mission as a response to the Announcement of Opportunity for the NASA Discovery Program. The author's proposal was supported by Lockheed and was one of 28 submitted to the discovery program. In February of 1995, LP became the first peer-reviewed and competitively selected Discovery Mission, was the least expensive by a factor of four and was also the only one to have been selected to proceed directly to construction, test, launch and flight.

The author, who was also the Program Manager and Mission Director, flew a near perfect 19 month mapping mission between January 7, 1998 (launch) and July 31, 1999 (End of Mission Impact at the Lunar South Pole) that produced global mapping data sets that were up to ten times better than proposed in the original Discovery Proposal. Thus LP was an unqualified engineering and scientific success and hence provides a reliable basis for estimating the minimum costs of successful commercial lunar missions, be they for science, entertainment, education or a combination thereof.

The total cost of the LP mission was $63 million for the nominal, one-year mission and $65 million for the entire mission with the following breakdown.

Spacecraft Bus	$20 million
Science Instruments	$04 million
Launch	$29 million
Nominal Mission Operations	$06 million
Extended Mission Operations	$02 million
Lockheed Fee	$04 million

There are several things to note before accepting the $63 million as the cost of an equivalent commercial Lunar Prospector class mission.

1) Though Lockheed—being one of the major aerospace companies—has a high overhead and is relatively expensive compared to what I expect for a smaller, more dedicated company such as LRI and LEI (the author's companies), the LP contract was done in the new "Faster, Better, Cheaper" NASA mode and as such, Lockheed did everything it could to keep the costs down. Also, since most of the subsystems and spacecraft components were purchased from external, smaller vendors, Lockheed's higher rates did not apply to the majority of the costs of the spacecraft. Thus I doubt the cost of the spacecraft bus was more that 10% higher that what a smaller company would have charged.

2) By Discovery rules, the cost of tracking LP with the NASA DSN (Deep Space Network) were not charged to the project. Though I am not sure anyone really knows the true cost of the DSN, it is around $1 million/month. So LP's tracking bill for the 19-month orbital mapping mission was probably something like $19 million. However, the nominal mission was 12 months long, after which NASA granted the funding for the 7-month long, low altitude mapping phase—or the extended phase—of the mission. Considering just the cost of the tracking for the nominal mission, the DSN charges were about $12 million.

3) The Lockheed Martin Athena II rocket plus the Star-37 Trans-Lunar Injection (TLI) stage used to launch LP cost $29 million ($25 million for the Athena and $4 million for the Star-37). Though LP might have been launched on a Taurus with a TLI stage for only $23 million, Lockheed would not allow me to do so. Also, the Taurus' capability was marginal for the LP launch, but assuming that it could have done the job, I could have saved $6 million. If LP had been somewhat (but not much) heavier, I would have had to use a Delta for the launch and that would have cost $65 million. Further, a cost estimate from Kistler indicates that an LP class mission could be launched to TLI for $17 million. Thus the launch cost for a LP class mission range from $17 to $65 million. However, since the Kistler launch vehicle has yet to fly (and may never fly) I have to assume the use of the Athena II launch vehicle for the basis of this cost estimate—at least until Kistler has a real vehicle. So the minimum LP launch cost remains at $29 million for this exercise.

4) The five LP science instruments were built for less than $4 million (actually $3.5 million), of which the Gamma-Ray Spectrometer (GRS), Neutron Spectrometer (NS) and Alpha-Particle Spectrometer (APS) were built at Los Alamos for $2.5 million. In great contrast, the NASA GRS for Mars Observer cost $20-million alone! The vast savings for the LP instruments were achieved by building instruments that had flight heritage and required essentially no development, e.g., the LP Magnetometer (MAG) and Electron Reflectometer (ER) were

duplicates of the Mars Observer and Mars Global Surveyor MAG/ER. Thus the cost of the LP science payload was minimum and certainly any non-science, entertainment payload (TV-camera, etc.) would cost a roughly the same.

5) Though the LP mission was dedicated to science, the power consumption, mass, data rates and operational requirements of the five-instrument payload were all at an absolute minimum, e.g., the downlink data rate was just 3.6 kbits/sec. This data rate is orders of magnitude lower than that required for an imaging mission, be it for science or entertainment. Thus, by design, LP's payload requirements were far below those of any possible follow-on mission, regardless of their nature and hence any follow-on mission (orbiter, lander or rover) will be much more costly than LP.

6) Lockheed's fee was restricted by government regulations to be no more than 10% of the mission's cost and was set at just $4 million or less than 7%. Since a commercial venture will want at least a 10% profit (if not much more), the fee on a PL class mission must be considered to be a minimum of 10%.

Given the above, my best estimate for a LP class commercial, one-year long, orbital mission for science, entertainment, etc. is $76 million with the breakdown as follows:

Spacecraft Bus	$18 million
Payload	$04 million
Athena II Launch	$29 million
Mission Operations	$06 million
Tracking	$12 million
Profit	$07 million

If a Delta were needed for the launch, the LP class mission would cost would be $116 million or nearly double that using an Athena II launch. If a Kistler launch were available, then the cost would be $63 million.

Thus there are two things to note from the above analysis. First, the $63 million cost of the real LP is truly representative of the absolute minimum cost of the simplest commercial lunar mission. Second, the current uncertainties in launch vehicle costs make a factor of two uncertainties in the cost of an LP class commercial mission. The cost estimates for more advanced lunar orbiter missions, lunar lander missions and lunar rover missions are based on extrapolations of the LP costs and are necessarily quite rough. This is the case because the number of undefined cost variables is large, at least until specific missions are developed to the Phase B Level—and that is currently not the case. For example, though some missions will be of short duration, e.g., a sample return mission lasts only a few weeks, their tracking and mission operations costs will be minimal; while others, such as advanced orbiters, rovers and seismic stations, will have tracking and operations costs equal to and even greatly exceeding those of LP. Also, as explained above, the costs of any more advanced spacecraft will be significantly more than that of LP and the launch costs will vary widely, depending on the mass of the spacecraft and the availability of new, inexpensive launch vehicles such as those being developed by such firms as Kistler and SpaceX.

Thus, the best that can be said at this point in time is that an average commercial lunar mission (minus launch costs) probably will cost 50% more than the cost estimated for an LP class commercial mission given above and the higher end commercial missions (again minus launch costs) will be at least twice (if not three times) as costly and the LP class mission. Given these rough extrapolations, the following gives the range of costs (to the nearest $5 million), including launch costs, for a range of missions and for low and high launch costs.

Launch Type	LP Class Mission	Average Mission	High-End Mission
New Space	$65 million	$85 million	>$110 million Delta
	$115 million	$140 million	>$160 million

The reader should remember that the above rough cost estimates are based on the assumption that there is only a 10% profit charged on these missions. If the investor base is unwilling to accept such a low return on investment and NASA (my assumed customer) is able to pay more than the current allowable 10% fee for commercial missions or if some other customers can be found for such missions, then the above cost estimates are lower by what ever level of profit the reader wants to assume.

Given the author's experience with the Lunar Prospector mission, the best estimate for the costs of commercial missions range from $65 million to at least $160 million, with the average costs running roughly at $100 million per mission. Thus any viable commercial program, such as the NASA Data Purchase Program being proposed by the author, needs to be able to support at least a couple of missions per year and therefore needs to be funded by the federal government at the $200 to $300 million level per year.

Editor's Note—Dr. Binder's numbers for the cost of launch at the time of this writing were actually pessimistic. As of the date of final edit, the cost quoted by Elon Musk of SpaceX for a launch that could handle all three variations is approximately $18 million, vs. the $29 million cited here. Through competition, this launch price and perhaps even lower costs throughout such a mission might be achievable.

LUNAR DATA PURCHASE PROGRAM RATIONAL

It is proposed that Congress build on the *Commercial Space Act of 1998* by appropriating $300 million per year for a *Lunar Data Purchase Program* and by directing NASA to implement the program.

There is ever growing support for the concept that a large fraction of the space exploration functions now carried out by NASA should be turned over to the commercial and/or private sectors. There are many reasons for this several yearlong trend and they include the following:

1) NASA programs are characterized by poor management, serious schedule slips and large cost overruns. The current prime example of which is the International Space Station—the ISS was originally budgeted at $8 billion and was to be operational in the early 90's; its current cost is now estimated to be from $40 to as high as $80 billion (the SSI was $4 billion over budget this FY alone) and it is still not fully operational;

2) the cost of failed missions is carried by the American tax payer with no return on investment—the current prime examples of which are the dual Mars failures in 1999 that cost the American tax payer over $300 million;

3) as reported in the national news media, NASA is now ranked as one of the four least competent and least productive of all the federal agencies; and

4) as recognition by essentially all parties, because of the competitive nature of the market place, the commercial and private sectors are more competent and more productive than federal agencies in general and much more so than NASA.

These issues provided Congress with part of the rational for its October 28, 1998 passage of the *Commercial Space Act of 1998* (Public Law 105-303), in which it is stated in Sec. 105 (a) that "The Administrator shall, to the extent possible and while satisfying the scientific or educational requirements of the National Aeronautics and Space Administration, and where appropriate, of other Federal agencies and scientific researchers, acquire, where cost effective, space science data from commercial provider".

Since the passage of the *Commercial Space Act of 1998*, the highly productive Lunar Prospector Orbital Mapping Mission successfully demonstrated that commercially based, very low cost lunar exploration missions can reliably produce high quality data, thereby demonstrating that such missions can be done as commercial ventures.

Together, the *Commercial Space Act of 1998* and the Lunar Prospector Mission provide the basis for Congress to take the next critical and logical step in the development of a viable, USA, efficient, commercial space exploration capability. That step is to start a NASA implemented *Lunar Data Purchase Program* with a funding level of $300 million per year. Such a *Lunar Data Purchase Program* would have the following benefits for the USA and its space effort:

1) The American tax payer would *only* pay for the data after is was delivered to NASA and verified in terms of its quantity and quality, thus the tax payer would not pay for failed commercial missions;

2) commercial missions would reduce the costs of space exploration, as shown by the Lunar Prospector Mission, by a factor of three to five, thereby increasing the return on investment for the American tax payer;

3) the $300 million/year funding level would provide a sufficiently high mission flight rate (two to three missions per year) so that investors could expect a reasonable return on investment, thereby ensuring that that a viable USA commercial lunar exploration capability would develop within a few years after its inception;

4) within several years, the *Lunar Data Purchase Program* would bring our knowledge of the Moon and its resources to the level where the business community could build a commercial lunar base and begin to utilize lunar resources for the benefit of humankind, e.g., to provide power to a power starved Earth;

5) once a commercially viable Lunar Exploration and Utilization Business Capability is at hand and investor/customer confidence in it has been firmly established, the newly developed business practices can be applied to the exploration of Mars and the rest of the solar system.

THE MOON'S RETURN ON INVESTMENT FOR HUMANITY

Based on our current knowledge of the Moon, the following is a list of potential benefits for humanity that may be derived from the Moon once even modest mining and industrial capabilities have been established there. Note that materials derived from the Moon can be delivered to Low Earth Orbit (LEO) at a cost that is less than few percent of that if the materials were launched from the Earth to LEO because of the Moon's low gravity.

Common Lunar Resources and their Uses:

- Abundant Sunlight – electrical power
- Cast Basalt – construction material for use on the Moon
- Oxygen – life support and rocket fuel
- Silicon – solar cells for generating solar power and alloying with iron
- Magnesium – construction and raw material for use on the Moon and in LEO
- Iron – construction material for use on the Moon
- Titanium – construction and raw material for use on the Moon and in LEO

Aluminum – construction and raw material for use on the Moon and in LEO
Nickel, Chromium, Manganese, Zirconium & Vanadium – alloying with iron for use on the Moon and raw material for use in LEO

The Moon can provide the Earth with Pollution-Free, Inexpensive Electrical Power. The value to Earth is estimated at $15 Trillion/Year

Lunar ^{3}He—the best fuel for the production of energy on Earth via nuclear fusion
Solar Power Satellites—built in Geostationary Orbit from lunar materials
Solar Power Farms—built directly on the Moon from lunar materials

Lunar Resource Construction Materials can serve Humanity in several Ways:

Construction of Solar Power Satellites – as above
Construction of LEO Manufacturing Facilities – space manufacturing of products for Earth made from raw lunar materials
Construction of Industrial, Habitation, Laboratory and Food Growth Facilities on the Moon – Provide the basis for an Earth/Moon economy, lunar colonization and research facilities for the study of the Moon, astronomy, physics, low-gravity biological and medical research and many other fields
Construction of Unmanned and Manned Spacecraft – expand humanities presence into the rest of the solar system and beyond

SUMMARY AND CONCLUSIONS

The US Space Program is now at a critical crossroad, either we alter the way we conduct our exploration of space and change it to a commercially based effort with a return on investment or we continue letting NASA lead us down the spiral towards collapse that began at the end of Apollo. Apollo was a politically driven, well-funded program with a definite goal NASA was given y the President. Since then, NASA has totally failed to provide the nation with the type of bold program that would lead to a return on investment and has failed with both of its two major, post-Apollo programs—the Shuttle and the International Space Station.

At the end of Apollo, NASA presented the USA with a forward thinking post-Apollo program, using upgraded Saturn 5's and advanced Apollo hardware that would have had a Lunar Base with a crew of 50 to 100 on the Moon by the early 80's, at the latest. Incapable of selling the President, Congress and the Nation on the concept, building on what we already had and leading to a return on investment, NASA entered the Shuttle and then the Space Station eras, both of which proved to be extremely costly in terms of dollars, schedule slips, failed objectives and, in the case of the shuttle, in human lives. Instead of leading the nation into the future, NASA has led us nowhere during the past here decades.

With the impetus of Columbia disaster, as well as the Space Station cost overruns and schedule slips, and in the absence of NASA providing any leadership, President Bush has wisely set the goal of returning man to the Moon and then going on to Mars. These are the same goals his father set some twenty years ago. But this time, the President has correctly added the commercial component to the mix.

As shown by Lunar Prospector, commercially based lunar missions are done at a cost that is 10 to 20% of that NASA would spend on the same missions and done on a shorter time scale and without massive schedule slips. Thus my experience leads me to believe that, in contrast to a doomed NASA effort, a commercially based Lunar Exploration and Lunar Base Program will succeed in A) leading the USA back to the Moon, building a commercial Lunar Base, B) developing an industrial capability to utilize lunar resources for the benefit of humanity and C) providing a return on investment.

Dr. Alan Binder is the founder and Director of the Lunar Research Institute (LRI), a non-profit, tax-exempt, science based corporation dedicated to the exploration of the Moon and its resources. Dr. Binder is also the founder and CEO of Lunar Exploration Inc. (LEI), a for-profit, engineering based corporation dedicated to building lunar spacecraft and conducting lunar exploration missions. He has 40+ years experience in the space program as a lunar and planetary research scientist and mission planner. Dr. Binder's research includes both planetary astronomy and geo-sciences as applied to planetary studies. Dr. Binder was a principal investigator on the 1976 Viking Mars Lander Program and spent 10 years teaching and doing research at various German universities (He speaks fluent German). He has also been active in advanced mission planning for both NASA and ESA in the areas of manned and unmanned lunar and planetary missions. Finally, Dr. Binder has authored and/or co-authored over 80 research papers during his career and has written a book, *Lunar Prospector: Against All Odds*, detailing the Lunar Prospector program.

If there should ever ba a PHD awarded for life experience in the cause of trying to develop commercial lunar activities, it would go to David Gump. Since the early 1990's he has worked tirelessly to put together projects that would cash in on the sex appeal of space for non-space clients. As a fellow co-Founder of LunaCorp (which David ran until recently) I watched him over the years as he was thrown out of one corporate board room after another. Not because his ideas didn't make sense, but usually because those to whom he was pitching couldn't "get it" when it came to space, which they, like most people had been indoctrinated to see as a government domain. Finally, after years of trying, he was able to interest a few visionary business people in testing the waters of space based advertising. This resulted in his producing the first ever commercial to be shot on the International Space Station, a warm and folksy spot for Radio Shack, featuring Russians as its stars (since he couldn't get the needed co-operation out of the American space agency.) Other successes followed, and now David is leading a consortium of New Space firms who are working with NASA (now under new leadership) to not only get us back to the Moon, but to make it pay.

A Real Return to the Moon

by David Gump

When the hatch opens on the next spaceship to touch down on the Moon, the people who step out won't all be NASA astronauts (nor even Chinese taikonauts, for that matter). Among them there is likely to be a winner of a global contest, financed by television networks and corporate sponsors. Every minute of that winner's adventure will be beamed live back to Earth, in a "Big Brother" 24/7 fishbowl experience. Or there could be a high-profile personality—a celebrity or network anchor—providing the human interest.

This outcome will become inevitable result of growing commercial activities in Earth orbit. For example, by about 2010, several competing private spacelines are likely to begin carrying people to low Earth orbit for less than $5 million each. By 2014, the cost of a ticket will be less than $2 million, and there will be small private habitats, satellite-assembly workshops and mini-warehouses in low Earth orbit. (By comparison, 2014 is the year when NASA hopes, if it gets full funding, to begin operating vehicles that carry about four people to space with a forecasted cost that could hit well over $25 million per person.) The advent of relatively cheap seats to orbit by private services in the 2010-14 period means that a commercially-based Moon landing is sure to follow a few years later.

The notion of affordable private spaceships may seem absurd, but the truly absurd idea is that space travel, alone among all high-tech industries, should always be expensive. Yuri Gagarin reached orbit when television sets had vacuum tubes and engineers designed rockets with slide rules and drafting paper. Now the average high school student owns a computer more powerful than any available to NASA during the Apollo program—with instant Web access to knowledge bases that would make 1960s space engineers moan with envy. The real question is not how could space travel become affordable, but how much longer can the relentless decline in the cost of all other technologies be kept from oozing into the aerospace arena?

By the way, those vacuum tubes are still hard at work on the Russian side of the space divide. I provided Lance Bass (of the band NSYNC) with all his start-up funding via a RadioShack sponsorship, and accompanied him during his initial medical screening at various sites around Moscow. Russian space facilities are the Land that Time Forgot, with ancient technicians still turning the worn dials on the original 1960s analog control panels of centrifuges and heart stress tables.

This is not to belittle the Russian space effort, but to emphasize that getting to orbit can be done perfectly well with used equipment, crumbling buildings and a work force that moonlights as cab drivers. Gleaming multi-billion dollar launch complexes and armies of quality assurance engineers aren't required; just competence and modest hardware.

In addition to pioneering with Gagarin, the Russians also created a breakthrough in passenger space travel by forcing NASA to accept commercial visitors to the International Space Station. This demonstrated that at least a small market exists for $20 million passenger tickets to orbit—and that ordinary folks with a few months training could journey off planet. This shattered the "imagination barrier" that reserved the role of space traveler to right-stuff fighter pilots.

On the U.S. side, we've believed that orbital space travel is so difficult that only the Federal government can marshal the money and the expertise to pull it off. This is something of a self-fulfilling prophecy. If we expect that space travel requires a cast of thousands and a budget of billions, that's what we'll get. The four Space Shuttle orbiters of the original fleet required about 4,000 workers each to achieve two or so flights per vehicle per year. By comparison, even the worst-managed airlines keep their planes flying two or three times per day with only about 100 workers each (this counts everyone, from ticket agents to pilots). To begin moving toward the airline level of reliability and low labor requirements, spaceships have to stop being science projects and become business tools.

And that's what's happening now that spaceships are being designed in the private sector: their owners are demanding minimal labor costs, simplicity in operations, and use of the most appropriate rather than the most advanced technologies.

What does it cost to create a passenger-carrying commercial spaceship? Looking at recent efforts, between $200 million and $500 million is my estimate. These are large amounts, but within the range of privately managed endeavors. The lower bound will deliver a mostly expendable vehicle with a two or three person capacity, and the higher end will produce greater reusability and a four to eight passenger capacity—and thus lower long-run per-passenger operating costs. Several variations on size and reusability could be commercial successes, just as small business jets, medium regional jets and huge jumbo jets co-exist in today's air travel industry. There won't be another "National Spaceship" like the NASA Shuttle ever again.

The leaders in 21st century space travel may be self-financed firms like Elon Musk's SpaceX and Jeff Bezos' Blue Origin. (Both made Internet fortunes: Musk with PayPal and Bezos with Amazon.com.) Some of the new private fleets will get their initial fuel from NASA or Pentagon contracts; the Defense Advanced Research Projects Agency is paying for the complete flight testing of at least one, and perhaps two, new low-cost small launch vehicles for cargo. Once flying, their designers can transition the vehicle to serve the much larger markets for commercial passengers. Or, the new space travel leaders may evolve from the X-Prize suborbital contestants, such as Burt Rutan's "SpaceShipOne" that was financed by Microsoft billionaire Paul Allen and may be commercialized by Sir Richard Branson's Virgin Galactic. But within the next five to six years, with development costs now so low, several of them will begin offering passage to orbit for less than $5 million per person.

Market research done by Derek Webber when he was at Futron Corp. showed the impact of these lower prices. Financed by NASA, the Futron study included detailed interviews with several hundred affluent people to determine their interest in personally traveling to orbit.

After an explanation of the hazards and discomforts of buying today's only existing commercial passage (including six months in Moscow learning Russian), they were asked how much they'd pay for a ticket to space. About seven percent said they'd buy a ticket regardless of the cost—going as high as $25 million. However, less than 1% of the sample actually was wealthy enough to buy a $25 million ticket. Most had net worths in the $1-$5 million range. This 7% buy-regardless-of-price can be tossed out of the responses to find the true demand curve—from the answers of wealthy people who actually stopped to consider the cost of the ticket. The chart below shows that the profound difference between today's $20 million price of a Russian ticket and the $2-$5 million tickets coming in three or four years. When the demand for $5 million tickets is matched against the supply of deca-millionaires able to buy them, one finds several hundred to several thousand tickets will be sold this decade.

Consider: several hundred millionaires traveling to orbit.

Most of them will be wondering: "This feels like the start of something big, the new space frontier that always seemed to be too far away to contemplate. But now I'm floating in zero gravity, looking down at the blue marble of Earth just like the ancient astronauts of Apollo. How can I invest my wealth to take advantage of this new era?"

And those thoughts, from hundreds of millionaires, will add more power to the Moon movement than any Congressional appropriation ever could.

They will build low-cost habitats that quickly will sustain more space explorers than the International Space Station. They will build orbital workshops, where satellites can be assembled for myriad task from communications to energy supply to remote sensing. They will build transportation centers, with warehouses, where specialized space vehicles can exchange cargo and people going to various destinations. The first incarnations of the low-cost habitats, workshops and transportation centers could be in operation by 2014 (about the same time that NASA hopes to fly its Crew Exploration Vehicle). The initial NASA week or two-long "expeditions" to the Moon—not a colony or sustained presence—are set for 2015 to 2020, if Congress actually provides the increased appropriations that NASA will require.

My bet is on the free enterprise crowd. With several hotels and workshops and transportation centers operating in low Earth orbit by 2014, networks and corporate marketers (and perhaps billionaires like Paul Allen and Richard Branson) likely will step up to participate. They may finance a free-enterprise Moon landing—selling tickets to NASA researchers, media companies and the (wealthy) public—or they may create a public-private partnership with NASA that enables a civil servants and members of the public to cooperate on an expedition.

Respondents Who Would Buy at Various Prices

	$1 m	$5 m	$10 m	$15 m	$20 m
Actual Futron Study Results	30%	20%	16%	9%	7%
After Dropping the 7% Who Buy Regardless of Price	23%	13%	9%	2%	0%

Source for "Actual" numbers:

NASA-funded "ASCENT Study Final Report" Jan. 31, 2003 by the Futron Corp.

(First space travel survey conducted among wealthy individuals - net worth more than $1 million)

Then what?

Learning the full story of what the Moon can do for humanity is a project that will take some years. The Apollo heroes only spent a total of one man-month on the surface, and didn't stray far from various sites near the equator. yet, later observations from orbital satellites discovered that the Moon's poles may be its most valuable locations, offering the potential for water and other resources.

Because of the low angle at which sunlight strikes them, the lunar poles also have attractive climates, compared to everywhere else on the Moon, which bake during two weeks of continuous sunlight and freeze through two weeks of darkness. The Apollo expeditions at the equator sidestepped this temperature battle by landing during "early morning" and blasting off long before the several days of "noon" pushed temperatures hotter than boiling water. (And there never was any attempt to stay through the night, when temperatures are colder than liquid nitrogen.) However, at the poles, habitats can be placed where sun and shadow meet, in a shelter of temperate conditions. (Various technologies also can exploit the sun/shadow temperature differential to generate electricity for the lunar settlers.)

Initial lunar commerce likely will be centered on science and entertainment, and then will branch out to mining and industrial activities. For scientists, the lunar far side is great for radio astronomy because it is shielded from the cacophony of emitters on Earth, and at night it's also shielded from the natural radio emissions of the Sun. Optical telescopes could be used to spot Earth-size planets around other stars.

The Moon also may reveal a two-billion-year record of what's happened in Earth's neighborhood, from periodic asteroid swarms to solar wind variations. With no rain or snow to erode the geologic record, we may be able to peel back the years to discover whether the Earth's neighborhood goes through regular bombardments, or how much the Sun's output has varied over the millennia. This could be vital in preparing for future bombardments, or coming fluctuations in the Sun's output.

In the first wave with the scientists will come the entertainment industry. Directors and cinematographers will make television and Web programming about the first lunar settlers, some of whom will be there in "gold rush mode" trying to be the first to discover economically useful ways to use the Moon's resources. Other programming will feature athletic contests in one-sixth gravity—with competitors perhaps drawn from the initial settler communities. In addition, the leading edge of adventure tourism—by the extremely wealthy—also will touch down on the Moon. It won't take many multi-millionaires and billionaires visiting to quickly outnumber the scientists.

The next stage likely will be mining and industrial activities, but no one can be sure at this point. After all, it took decades of Western pioneers skipping over the Great Plains to reach California before folks realized that vast quantities of grain could be grown on what initially was called the Great American Desert of Kansas, Nebraska and the Dakotas. New technology helped; railroads made routine transportation affordable even without the navigable rivers found along the coasts; and steam power meant grain could be milled without the water-driven grinding mills that flourished along coastal rivers. Inventive minds will gradually discover the Moon's true resources, based on the technologies and economics of the mid 21st century.

For example, asteroid impacts on the Moon may have left certain craters rich in platinum-group metals. On Earth, entire mountains are stripped down to find a few tons of platinum, palladium and related metals that go into such essentials as pollution-control devices for cars. An ounce of platinum sells for more than $1,000, so a ton of it would be worth $32 million back on Earth. But would it ever make economic sense to extract the metal from the crater, and then move that ton from the Moon back to Earth's factories?

The answer depends on clever engineers exploiting the new economics of the Moon-Earth system. For example, ore processing would take place in a vacuum. This actually can greatly simplify the process; without oxygen around to instantly corrupt metals (oxygen plus iron equals rust!), metal-rich ores can undergo a straightforward baking process to boil off and individually collect distinct metals just like an oil refinery splits crude oil into its various components of gasoline, diesel, etc.

The next step of transportation also can take advantage of the Moon's special properties. Over most of the Moon, there's constant strong sunlight (no clouds) for two weeks at a time for the creation of electricity. This can be fed into mass drivers, which would magnetically accelerate small chunks of metal into lunar orbit. Once shot out of the mass drivers, there's no atmosphere to slow the packages down until they are caught by a yet-to-be-designed device in lunar orbit, or at a Libration point. (There is a Libration or Lagrange point in front of, and behind, the Moon where the gravity of the Moon and the Earth balance each other out, and stations or ships can stay virtually indefinitely without using much propulsion. Three other Libration points exist at various locations on the Moon's orbital path.)

The return trip from the Moon to the Earth has two advantages over the outbound leg. Lifting off from the Moon takes only 5% of the energy required to launch from the Earth's surface, and with mass drivers inert cargo like metals can be hurled without using any expensive chemical propellants. (The acceleration reaches several hundred gravities, so it's not a survivable route for humans, unfortunately.)

And once accelerated on their way toward Earth, the metal shipments just keep on going. There's no friction to slow them down, unlike the continuous force required to keep planes, trains and automobiles moving on Earth. Space travel is not really about distances so much as it is about velocity. Once a ship achieves a high velocity, it basically coasts the rest of the way to its destination. With lunar exports, after a small kick to get off its surface, cargo basically coasts "downhill" toward the Earth's gravity well.

The other great thing about the return trip is that the kinetic energy of the cargo coming down to Earth from space can be sold to ships ascending from Earth's surface. Cargo shipments "caught" by Earth-orbit transportation centers (using miles-long tethers) will give up their velocity to the centers, raising their orbits. Ships can take off from Earth with partial fuel loads, and complete their trip by catching tethers hanging from the centers. This transfers the missing final burst of energy and velocity to the ships so they can reach orbit, and in the process brings the centers' orbits back down a bit. In essence, instead of firing retro-rockets to slow down the lunar shipments so they de-orbit into the Earth's atmosphere, their kinetic energy is conserved by transferring it to the transportation centers. (Ships returning to Earth can do the same thing, using tethers to decelerate instead of killing off their velocity with retro-rockets.)

Cargo shipments from the Moon—whether they turn out to be platinum ingots for pollution control devices or something else—thus have two ways to pay their way: their sales value back on the home planet, and their kinetic energy that can be sold to cargo and passengers trying to lift off from Earth.

If a lunar export trade still seems unlikely, consider what's happening right now on Earth. Trash pickers across America are scooping up metal junk, loading it on to rail cars for shipment to ocean ports, where it is transferred to freighters for a 10,000-mile journey to China. Then this trash is smelted back into steel, rolled into sheets and assembled into microwaves and washing machines that are shipped right back to Best Buys and Sears stores across the country. If something as improbable at this cycle makes economic sense, then lobbing cargo into a downhill coast from the Moon to Earth may make sense too.

The lunar settlements also could export pure energy, and skip the transportation issues. Solar power cells could be made directly from lunar elements, and vast electricity farms could generate power that's beamed back to Earth via microwave or laser. The rectennas that collect this energy on Earth might be co-located with wind turbine fields, taking advantage of the existing dedicated surface area and the pre-existing connections to regional power grids.

Of course, today's forecasts for the Moon are like 16th century experts sitting comfortably in London or Madrid, speculating about what wealth the newly discovered Americas would produce. The answers came only after centuries of exploration and innovation (plus genocide, slavery and several wars). The Moon and Mars frontiers have no indigenous peoples to displace, and the Moon, at least, definitely lacks any ecology to disturb. We can have big dreams now, with a shot at achieving them peacefully and for the good of all humanity. (30)

David Gump, president of Transformational Space Corporation LLC in Reston, Va., has worked for 15 years to promote commercial involvement in Moon exploration, with presentations to more than a hundred mainstream U.S. corporations in many industries. As president of LunaCorp, he directed projects on the International Space Station ranging from demonstration of RadioShack products to having Major League Baseball and Fox Sports feature ISS during the 2002 World Series. He also arranged for all of Lance Bass's initial funding and was involved in the negotiations with networks and major sponsors. He previously founded the newsletter "Space Business News" on commercial use of space, and he organized two national conferences on business use of the Space Shuttle and the ISS. He is the author of "Space Enterprise: Beyond NASA" published by Praeger Books in 1990.

The central challenge facing our nation as we begin the quest to return to the Moon is that of cost, and the main driver of costs in space is transportation—primarily the cost of the first couple of hundred miles as one lifts out of Earth's gravity well. As I mentioned in the preface, unless our government gets out of the Earth to LEO transportation business and learns to buy its rides to space as part of a larger market it will not be able to afford to keep people on the Moon, Mars or anywhere else. This does not mean the private sector will not do so on its own, but without the large amounts of cash the government routinely throws at (and usually wastes) on transportation systems, it may well be slow going. Wouldn't it be great if both parties were being mutually supportive? Dr. Lurio has been fighting this fight along with myself and others in our movement for well over a decade. His logic, cool rational thinking and sheer doggedness when engaged in a fight for this or that element of change in the space transportation arena have earned him the respect of many in the New Space movement and the space related media. In the following piece he greatly expands on my comments at the beginning of this book, and clearly lays out many of the elements needed to revolutionize human access to space.

The New Space Revolution and Return to the Moon

by Charles Lurio, PhD

In the parlance of engineering, the "long pole in the tent" is that part of a task that paces the objective. For any human activity in earth orbit and beyond—commercial or non commercial—the 'longest' pole is the price and reliability of transportation to Low Earth Orbit (LEO); just behind are the skills necessary to maintaining humans in space. One cannot move these capabilities to practicality by relying only upon the historically small volume and specialized use needs of government space efforts. Mutually competing New Space firms with experience in commercial markets many times larger than today's are essential to achieving prices consistent with a sustainable human presence on the Moon and beyond.

The main obstacle to achieving success is not technology. Numerous schemes have been devised for renewed human space activity beyond LEO since the Apollo program, yet all have foundered upon the rocks of excessive cost and/or politically unsustainable timelines. The inability of NASA to dramatically mitigate these problems is not a fantasy of those instinctively hostile to human space exploration: it is reflected by the Shuttle and ISS experiences.

NASA's approach has been to create unique engineering systems which in whole and in part incorporate none of the practical real world experience forged by significant demand and competition. To expect affordable cost and tolerable safety from an Earth- Moon transportation system in support of a Lunar community is as ludicrous as expecting practical sustenance of Antarctic research stations minus any previous experience with boat building and sailing. The problem is the lack of "Economics 101," not of technology or of public funding. What should have been the exception—central planning for a singular goal—has in practice, if not law, barred normal, trial-and-error ventures into potentially profitable new space-related markets since Apollo. If central planning were the path to practical and profitable technologies, the Soviet Union would have become the font of consumer goods for the world.

Apollo was a specialized cold war competition that in its time was seen as a near-incredible achievement. The unintended result was to embed the belief that fundamentally the free market and spaceflight were virtually incompatible. This manifested itself in two major ways: First, the space establishment became wedded to the absurd notion that the expense of engineering components divorced from market forces was the 'natural approach' to space exploration and that even if the resulting delays were indefinite, they were unavoidable; Second—and worse—the public, the business community and the Congress came to see NASA and its space establishment as the 'Tsar of Spaceflight' for anything other than comsats. Thus, a willing audience has been created for the consequences of NASA's own delusions, most glaringly that practical human activity in space has to start with vast expense and add ever more outlandish technologies rather than use sustainable markets.

Contrasting circumstances in computer systems since the 1960's nurtured very different views. While as late as 1977, Ken Olson—President, Chairman and Founder of Digital Equipment Corporation—could smugly assert that, "There is no reason anyone would want a computer in their home," his was only one of many voices, and there was no 'Tsar' of computer systems hobbling the national imagination. Thus, free market forces were able to spark multiple paths for creative engineering, and incremental—but cumulatively rapid—improvements could result. Thus computers evolved from remote, costly and somewhat mysterious machinery to ubiquitous personal devices.

Only if we are freed from the spurious attitudes of the Apollo 'hangover' can we have analogous improvements in the possibilities and practicalities of spaceflight, including human space exploration. The right set of national policies may be able to ease the way for expansion of the New SpaceNew Space firms' capabilities into a widening set of new markets, starting with LEO access. The resulting practicalities in cost and operations can lead to an achievable human exploration effort.

Passage of The Commercial Space Launch Amendments Act of 2004 (HR 5382) in the last moments of the 108th Congress was an important step. It sets an essential precedent by creating a basic law for suborbital human spaceflight markets that avoids crippling safety restrictions on development and operations. Although it has its imperfections; its most important attribute is that it reflects a respectful dialogue between government and the prospective New Space industry.

In terms of tangible results there is certainly far yet to go: 'Traditional' government run space programs have used up many hundreds of billions of public dollars over the past 40-odd years. Yet the lowest cost price to orbit for the private citizen today is $20 million, riding on the same basic rocket that carried Gagarin in 1961. This is reason enough to look towards the radical and fundamental changes that New Space efforts can provide.

Requirements for New Markets, New Firms and Industries

To create fundamental change, New Space must exploit new markets that have unit demands that are orders of magnitude larger than those for present day comsats and government space operations. The most promising of these appears to be space tourism, starting with suborbital flights. It also provides an illustration of the characteristics that to one degree or another must characterize all viable New Space markets:

1) Incremental Technical Steps: The needed vehicles perform short suborbital 'hops', reaching space altitude but not going into orbit. These provide passengers with 3-7 minutes of zero-g experience and an impressive view while making only a modest technical demand upon the small, innovative new launch system developers.

2) 'Reasonable' Entry Cost: Suborbital system requirements keep investor entry cost for a Reusable Launch Vehicle (RLV) in the low tens of millions of dollars. Paul Allen of Microsoft fame spent an amount in this range for development of the "SpaceShip One" piloted X Prize vehicle by Burt Rutan's 'Scaled Composites.' Other wealthy 'pioneer investors' are looking at concepts such as XCOR Aerospace's 'Xerus' operational vehicle concept or are self-funding entire organizations to consider serving suborbital markets. This is true of the 'Blue Origin' research group financed by Jeff Bezos, founder of Amazon.com.

3) Achievable Operational Cost and Reliability Path: Successful Reusable Launch Vehicle (RLV) development must be guided by using the most robust, simple systems possible for a task and accumulating large amounts of hardware test experience. This rapid learning style was proven in the development of many pioneering aircraft and was applied to 'SpaceShipOne'. It allows confidence won from well tested, robust systems to replace expensively 'stacked' backup systems and bureaucracies. Unambitious tests can be relatively cheap and numerous; as they become more challenging and expensive they are fewer, but stand upon substantive experience. Development failures are not avoided at all cost, but the program tries to ferret them out when the cost of change is minimal.

4) Initial Market Interest: Interviews by Zogby (on behalf of the Futron Corporation) with individuals having income higher than $250,000/yr. and/or net worth over $1 million showed nearly 20 percent having definite or very great interest in taking a suborbital flight at a price of $100,000. Conservatively, this may represent 200 individuals per year 'going suborbital'. Judging by interest in Branson's proposed 'Virgin Galactic' suborbital flights at $200,000 per person, demand may be far higher, though to be sure we shall have to wait until tangible customer money must be handed over.

5) Sustainable Market Interest: A business plan requires a minimum number of years to make sense. For the suborbital market forecasts over time look promising. Under even modest suborbital demand estimates the likely market is still vastly larger than that for any previous launch systems, creating an unprecedented forcing function for space vehicles toward lower cost and ticket price, operational ease and increased reliability and safety. The 'learning curve' is put into dramatic action.

6) A reasonable legal/regulatory regime: This is at least as important as the items above. Attempting to set passenger safety levels now without market experience could indeed guarantee perfect safety— but only because commercial flights could not go forward at all. We must ensure protection of those uninvolved with the flight; but equally, passengers must 'fly at their own risk' at least for some period, as allowed by HR 5382. This and similar provisions for an 'infant industry' are essential to avoid scaring off investors by making development and commercial payoff impossible. The highest safety can only be developed if the widest possible range of technical ideas have a fair shot at development. Conversely, to try to evade any regulatory framework and dialogue with government only amplifies the danger of the inevitable accidents causing an industry-destroying backlash of overregulation.

Once a suborbital passenger industry is in place, the technical base can be a start point to extend low cost and high reliability to orbital space tourism It is quite possible—though not a certainty—that revenue from the suborbital market alone will be sufficient to quickly stimulate private investment in practical, low operations cost vehicles for this purpose.

The potential for orbital tourism is supported by surveys and studies over the past several years (see, e.g. any number of items in the archives of www.spacefuture.com) that have shown that a large fraction of the public in developed countries would pay several months' income for a 'vacation' in space; many would pay more than one year's worth. Indeed, the interest in tourism to orbit seems much higher than that for brief suborbital flights. Even with discounting of demand for transition from an abstraction to a real prospect, surveys suggest that orbital tourism may lead to a business worth many billions of dollars per year.

A small market for piloted suborbital vehicles may be launching small satellites by releasing an upper stage in flight. Another market that has often been suggested is to provide fast package/human transport around the globe on suborbital trajectories. However, that is a less 'natural' market than it first appears. There are significant technical challenges for the vehicle versus those for presently planned suborbital systems. Moreover, it must compete head-on with the highly developed and proven existing conventional aircraft system.

Expanding Markets and New Skills

The development of orbital tourism should also result in practicality and far lower cost in a range of basic space habitability skills in addition to transport to LEO. (All of which have Lunar applications as well.) For example:

1. 'Housing' (i.e. for orbital hotels)—The extreme cost of International Space Station (ISS) modules is prohibitive for a tourism market. Far lower price and more effective systems—the most robust but least complex capable of doing the job for the market—will be needed. These may include the use of inflatable modules, of near-total environmental recycling and of in-space assembly of module subsystems. Bigelow Aerospace is presently investing very large amounts into development of inflatable modules that could be the basis of an initial 'space hotel'.

2. Space Construction and Architectures—To cater to tourist interest, new types of architectures for space habitation will be created. Large, open pressurized spaces for zero-g games, big windows and rotation-created gravity may be the route to larger profits. An analogy is to the amenities provided by recent generations of cruise ships. Many of these new options will require expanding construction abilities beyond simple docking to those requiring extensive Extra-Vehicular Activity (EVA).

3. Practical Pressure suits and Extra-Vehicular Activity (EVA) skills. Market forces thus could spur development of practical space operations and construction abilities. Prominently, this includes 'practical' pressure suits, whose development has been stalled by decades of centrally planned and minuscule government-run human spaceflight 'markets.' Present pressure suits are extraordinarily costly and extremely difficult to use. The wearer is rapidly exhausted by fighting low internal pressures. These low pressures (and thus high oxygen partial pressures), also require users to engage in hours of pre-breathing to prevent the 'bends.'

In addition to encouragement from construction needs, pressure suit innovation could also be driven by demand to provide space tourists with an 'EVA experience.'

While space may not be a benign environment, for many space activities dangers can be anticipated and needed reaction times made relatively long. Given low transport costs and reliable technologies, many margins of safety there can eventually be closer to those of civil engineering than of commercial aircraft today.

A New Space - Human Exploration Dynamic

The first benefit of New Space to human exploration is in transportation to LEO. If government is not hopelessly wedded to fiscally unsustainable, nut-and-bolt specified, limited use launchers, it can take advantage of the lower prices of systems derived from much larger private markets. The resulting greater mass per dollar of components that can be brought to orbit creates options for more affordable and safer human space exploration.

Other skills driven to practicality by New Space markets are applicable in whole or in part to Lunar/Mars transport, habitation and operations. 'Off the shelf' technologies in EVA, habitation, and support systems benefiting from free market learning curves will provide much lower cost and far higher reliability than unique systems developed under stultifying central planning habits.

Basic architectures of human space exploration may change as well. A useful option for long Mars voyages would be a practical pseudo-g rotating habitat and/or one large enough to reduce 'cabin fever.' Demands on transit time and propulsion system technology would be reduced. Interplanetary radiation levels would remain a problem, but more solutions for that are also made available. Reduced cost of transport to LEO would allow greater shielding mass from Earth directly, or as other transport and operations prices become low enough Lunar regolith can be exported for shielding.

In a complementary way, the development of specialized capabilities for government funded exploration can help expand the scope of commercial operations. Subsystems developed for exploration (but which have prohibitive initial research costs for New Space firms) may be used for cost reductions and increased safety of commercial space transportation systems, just as 'fly-by-wire', derived from fighter jets, has found its way into commercial airliners. Pressure suit adaptations and ancillary systems needed for 1/6th gravity, dust-laden Lunar conditions could accelerate firms' timetables for expansion of commercial operations to the Lunar surface.

An expanding circle of economic benefit, technical improvement and scientific discovery can be created between private space markets and public expeditions of exploration.

Implementation: Troubled Precedents

Present U.S. government human space exploration and space transportation plans speak of reducing costs by such means as 'spiraling' projects to stretch development demands and purchasing at least some private transportation to orbit. But the former does not change fundamental axioms while the latter is presently only a tepid commitment. The net result of these and other aspects of the exploration plan is to create another attempt at an 'Apollo rewrite,' which has always wrecked human space exploration proposals since that project.

For example, many in the space establishment are not even questioning the 'need' for low use rate, costly heavy lift systems to transport the bulk of parts to orbit; most such schemes rely on Shuttle components with their inherently high support costs. As in Apollo, NASA wants its own vehicle to carry out the task of astronaut access to LEO—as well as being the core for Lunar/Mars exploration. Yet for the period envisioned, private market driven human access to LEO is foreseeable at far lower cost. As well, even before extensive private markets develop in Earth-Moon space, the efficiencies that could be brought by the lean new space development firms could significantly reduce the cost of exploration-oriented systems.

Nor is the NASA record encouraging in self-described attempts to create a 'partnership' of public and private needs for space. The government might try to 'assist' the transition of New Space from suborbital to orbital human transport. The objective would be to allow companies to tap into the commercial orbital markets sooner and consequently also reduce cost more quickly of human and unmanned transport for the exploration program. Yet recall that government needs and private markets were supposed to meld in the 'VentureStar' follow-on to the X-33 test vehicle. NASA insisted on imposing fundamental parameters such as vehicle payload capabilities that prohibited a rational business case, rather than adapting itself to the latter. It (and the contractor) also celebrated rather than avoided the temptation to require development and testing of several excessively complex and difficult technologies in parallel for the vehicle to succeed. The X-33 predictably became a costly and notorious dead end rather than an incremental testbed pointing to a practical orbital system.

Traditional NASA contractors instinctively prefer the certainty of government contracts—present and future—over standing firm for the needs of commercial, albeit uncertain, outside markets and realism about the practical technologies they require. Fortunately, the psychology of the small firms entering the suborbital rocket arena is not that of government reliant corporate bureaucracies but of 'dot-com' venture capitalists; they have the prospect of large suborbital/orbital private markets, and lower entry investment cost than previous private space transport proposals. However, ham handed government attempts to 'aid' them or contracts requiring huge work-loads for documentation could thwart their ability to apply their inherent efficiencies and market driven cost advantages to a publicly funded exploration project.

Certainly the rate of future progress into the new New Space markets is uncertain. But if regulatory frameworks do not hinder them excessively, the prospects of their rapidly improving the cost and safety of LEO access is a far better bet than the failed path of central planning. Qualitatively different proposals for government to aid the progress of the private sector do have the potential for doing good rather than harm—if implemented properly.

But if past NASA planning rigidity rolls forward, the interaction between the agency and the new private space industry should not go beyond low cost technology programs under efficient contract rules—to protect the new industry from damage.

Policy Recommendations

The civilian 707 jetliner resulted from the Air Force's need for a military transport/tanker. Boeing invested its own R&D dollars knowing that the government's needs could provide the critical learning curve for a near-identical civilian aircraft. Were the annual and ongoing demand from of a human space exploration program large enough, it could similarly improve the prospectus for private investment in uncertain but potentially far larger non-government markets.

If the proposed exploration program is locked into old failed frameworks, it will project tens of billions of dollars for development and operation of unique systems for transport to LEO and operations there and beyond. Of course, the new private human space access industries could move forward anyway, assuming that reasonable 'infant industry' regulation is in place.

But by breaking out of Apollo-style rigidities and avoiding the mistakes of past NASA attempts to 'aide' the private sector, many billions of dollars can be saved in human exploration proposals. The following could be among the important strategies:

1. *Do not develop a Heavy Lift Launcher (HLLV)* An HLLV requires very large taxpayer investment up front, then keeps flight rates low, thus eliminating learning curves, keeping high launch cost and reducing resources for actual exploration.

Projected price to develop a 'clean sheet' 100 metric ton (i.e., Saturn V class) booster may run as high as $13-16 billion. Some state that putting together a heavy lifter from Shuttle-derived hardware might run to 'only' $3-4 Billion. Given the current standing army of shuttle support personnel, one suspects that operations costs using shuttle components would be higher than for the 'clean sheet' HLLV. In any case the message is that unless and until substantial private needs would pay for it anyway, an HLLV is a huge waste of public funds.

2. *Instead, Require total out-contracting of Lift to LEO, and be flexible in payload sizing and other specifications.* Use the learning curve rather than fight against it. Use commercial ELV lifters but plan from the start to include privately developed RLV (or near- fully reusable) systems that will incur ever lower cost per flight as private markets grow and apply genuine learning curves to them. Contracts for the exploration program's needs should be simple: perhaps a certain amount of cargo to LEO for a few years with only the most basic needs specified. A single page document should be the goal. Plans must also allow for breaking exploration payloads into smaller units than conventionally assumed. Let's consider the effects of these measures assuming an annual demand by an exploration project for 250-300 tons/year in LEO.

a. *ELV Systems.* Current large commercial ELVs can provide lift in about the 10-25 ton range (i.e., 10-30 ELV flights/year). If NASA were forced to lift exploration program components using packages in this range, it would avoid the launcher development cost and take maximum advantage of existing ELV launch price savings. And with 5-10 flights per year, the effect of the learning curve on the cost of an ELV begins to take hold. Two ELV suppliers could simultaneously and profitably provide this transport to LEO at a price in the low thousands of dollars per pound.

b. *RLV Systems.* For an RLV system, flight rates to create a viable business plan attracting private development dollars need to be perhaps 20-40 per supplier per year, sustained for a decade. With the same payload size as for the ELV case, the exploration program is inadequate to maintain the vital competition of having more than one supplier. (Neither a single supplier nor a single customer situation extracts the best quality and price from a market.) Still, a relatively low government demand might be able to 'top off' investor confidence at the beginning of private projects otherwise contingent entirely on new and uncertain markets.

But the leverage of Moon-Mars exploration can be used more effectively. If it is provided exclusively from Earth, fuel represents about 75% of the needed mass to LEO. Obviously it can be launched in smaller packages more easily than can hardware; it can then be stored in an orbital fuel depot. This could generate demand for several hundred flights per year of an RLV. That in itself would be more than enough to meet the business criteria for private development of several RLVs. And it would be far more effective in 'topping off' investor confidence as suggested above.

3. *Encourage, rather than discourage, competition for selected high technology developments among NASA Centers, focusing where private demand is not yet a driving force.* Over a period of years NASA policy changed from encouraging its Centers to compete amongst each other to develop particular technologies to concentrating their development at single Centers—in the name of 'efficiency.' Of course that was nonsense, since the only true efficiency of result derives from competition. This simply forced the failures of central planning to be the internal policy of the Agency. Decisively doing the opposite will not only benefit NASA's exploration needs but also improve the Agency's ability to create practical technologies able to accelerate expansion of the new private space sectors.

4. *Some other broad strategies to speed the development of privately financed Orbital RLV's and other technologies and applications with utility in the exploration program: a. Use Prizes, intelligently.* Prizes to stimulate innovation and achievement in space capabilities should be used with enthusiasm and intelligence. Bigelow is offering $50 million for a privately developed, at least 80% reusable system to carry passengers privately to orbit by 2010. The NASA 'Centennial Prizes' program has been created but Congress, NASA and the Administration seem wary of making it large enough to significantly affect the exploration plan. The larger the goal, and thus the higher the prize money, the closer the amount of prize money has to come to the actual cost of achieving the goal. Beyond a certain point there are

simply not going to be enough investors willing risk spending above—or even to— the value of the prize in order to win it. Paul Allen spent over twice the value of the $10 million X-prize for SpaceShip One, and he appears to have been the only entrant to have spent more (or even close to) the value of the prize in seeking it. In effect, he may have defined a rough limit beyond which, if you asked for more 'result' for the same prize money, there would be zero investors.

This may be mitigated somewhat if a commercial success is seen to result quickly from the X-Prize, which after all is the first of its kind for the space arena. Tangible steps to make prizes more effective not only include raising the prize value closer to the cost (as anticipated cost rises), but allowing for second and third prizes, so investors are not taking an all-or-nothing gamble. To reflect the standing of the competition results, the runner-up prizes would gradually go down in value with 'win, place, or show'.

A prime asset of certain prizes is the ability to give a psychological 'shock,' to change public (and investors') perceptions of where the limits of the possible lay in a given area of technical or market endeavor. The latter has certainly been an *initial* result with the X-Prize win. But of course, whether this is a *persistent* result will depend critically upon how actual markets pan out and actual regulations are implemented.

b. 'Genuine' X-vehicles A very successful lesson from aviation was the use of testbed vehicles funded by government for trying out ideas with military or civilian applications, as shown by the DoD and by NACA (NASA's predecessor). Such vehicles are not intended to show particular payload capabilities, or demonstrate radically new technologies all at once. Rather, they challenge the limits of what engineers can do by integrating existing or near-at-hand technologies, given budget and time limits. But without commitment and time to impose vast and fundamental reform there, it would be rash to let NASA have anything to do with such efforts. The prime cautionary tale is what happened DoD's proposed successor to the DC-X program, the 'SX-2'. When moved to NASA, it became that over specified vehicle and high tech development disaster previously noted—the 'X-33'.

Concluding Thoughts

In the past, prospective private space entrepreneurs have looked fruitlessly to the government as an 'anchor' customer in getting their plans off the ground. In parallel, human exploration proposals have been proposed and an equal number have vanished since Apollo. The difference today is that there is the prospect of an evolving set of private human spaceflight industries proceeding from sub-orbit to orbit and beyond to the Moon and perhaps someday even Mars, where government demand would be but a small part of the markets and could adapt to benefit from them. If the structural flaws of past government attempts to 'aid' the private space sector can be avoided, there is also an opportunity for government to accelerate initial progress of private human spaceflight industries. But given past experience it is uncertain whether the chance of a 'positive' intervention is worth the risk to the new private markets.

In any case, for the proposed exploration vision to be practical it must incorporate radical change beyond today's version. Adaptability must be permitted for how one provides payloads to orbit, for LEO operations and how one fulfills other goals of the exploration project. This flexibility is no more than that required every day of any company that wants to minimize costs and maximize profits in the free market. The difference is that instead of maximizing profits one wishes to maximize the ability to carry out the most extensive, extendable and sustainable program of human exploration.

The resistance to change from an Apollo style program contingent on detailed, long-range central planning is fearsomely embedded in structures of perception and in institutions. As it is said that generals always plan for the last war, so since Apollo NASA has always planned for the next Apollo.

The contortions in such plans have included to greater or lesser degree: huge price projections; development of fundamental capabilities stretched out over an unrealistic number of Congresses and Administrations; bizarrely overestimated requirements for technical complexity; a compulsion to use derivatives of 'existing hardware' no matter how impractically costly. Many of these have resulted from an institutional imperative to 'service' the NASA bureaucracy.

The exceptional cold war competition with the USSR created a perception that the supposed efficiencies of 'rational' planning with its appeal to 'neatness' were inherently needed for space achievements. Of course they are actually counterproductive, since they cannot appeal to market driven feedbacks. The latter produce engineering success defined as finding a workable—not ideal—path to utility in the real world.

The reforms needed to reap benefits from the New Space sector can aid a permanent return to the Moon in the most astonishing way of all: let it actually happen.

Charles A. Lurio obtained his Bachelor's, Master's and Ph.D. degrees from the Department of Aeronautics and Astronautics of the Massachusetts Institute of Technology. For several years after completing his academic degrees in 1988, he worked on NASA and DoD projects at several firms in Massachusetts and New Hampshire. After expiration of a DoD contract lead to a layoff, he began to spend more time on his long time interest in space policy, in parallel with performing consulting in both technical and non-technical areas. During a stint as director of the 'Cheap Access to Space' project of the Space Frontier Foundation, he became familiar with reporters based at several major media organizations. This experience and others reinforced his growing conviction that only by going outside of the 'space establishment' could any fundamental positive change be made towards a future for practical spaceflight. Since the end of 2000, he has devoted nearly full time to acting upon this conviction in several arenas. These include: creating contacts on Capitol Hill; writing, including guest editorials in Aviation Week and Space Technology and Space News; greatly expanded media contacts; and a regular set of updates and commentaries emailed to these contacts. He has become a trusted background and discussion source for major media, such as 'The Economist.' He has been cited or quoted in many publications including 'The New York Times' and 'The Wall Street Journal.' His most recent efforts have focussed on the need for stable regulatory frameworks for private human spaceflight.

One of the biggest debates occurring as part of our decision to return to the Moon is what the transportation infrastructure will look like, and who will operate it. Do we build giant heavy lift rocket systems, do we parse out the needed infrastructure payloads to smaller carriers. Do we throw large pre-constructed facilities from the Earth to the Moon in one shot, or do we set up ferry systems, with the equivalents of short haul carriers, tugs, long haul carriers etc. In this book you will read several technological alternatives, and several combinations of those alternatives. The one common theme is that whatever systems are used, if we are to succeed in creating permanent new communities off the Earth, they should be privately operated where ever possible—starting with the Earth to Low Earth Orbit segment of the trip. Mr. Strickland lays out his case in detail, as do other authors in this collection. Unfortunately (lip service aside) NASA still hasn't figured this out, and is looking to pour billions of dollars into re-engineering the space shuttle's launch system rather than trying to seed and feed a new commercial space transportation industry. But the debate continues, with groups like the Space Frontier Foundation, the Space Access Society and others trying to make the common sense case for a new approach. Perhaps by the time you read this all will be decided. More likely it will not. In that case, I offer you more food for thought.

ACCESS TO LUNA

Problems and Solutions for an Earth-Moon Transportation System

by John K. Strickland, Jr.

In accepting the Charter to the President's Space Exploration Commission, NASA states that it will "*Implement a sustained and affordable human and robotic program to explore the Solar System and beyond (and) extend human presence across the solar system, starting with a return to the Moon by the year 2020 ...*"

To create such a sustained and affordable program will take some very difficult planning, decision making, and changes in thinking, so that space operations are carried out in a very different manner compared to today. Fundamental to sustainability is reducing the cost of each discrete mission and the annual cost of supporting each installation or base. Much of this depends on making the exploration and supply trips economical, safe and reliable. To do this, we need to create a semi-permanent, re-usable transportation system between the Earth's surface, Earth orbit, and the Moon.

The first requirement is to get crew and lunar hardware into Earth orbit. A debate is now underway, within and without NASA, on whether to use existing expendable rockets for the Lunar phase of the Exploration programs or build new ones. Since the first use of any new booster (other than possible use for Space station support) is ten years away, there is plenty of time to design and build new rockets, but the decision to do so or not still needs to be made fairly soon. The desirability of spending money up front on a newer, cheaper space transportation system increases with the total tonnage of equipment and vehicles required per year, and with the duration of the program. The Lunar Program implied by the Charter is a robust program and involves more than just temporary exploration bases and limited scientific studies. However, decisions on launchers are likely to be made more on the basis of current budget pressures, and less on long-term program requirements and costs. Some studies are assuming minimum requirements comparable to an "Apollo 11" style program, which tilts the balance towards using expendables. To make the best decision on transport requirements thus requires a full assessment of mission and base requirements, and a starting assumption that they will not be temporary. The current shuttle system makes a good example of program longevity, and since it will have been in use for roughly 25 years (not counting the 2 stand-down periods after accidents), by the time it is retired between 2010 and 2012, we should use 25 years as a minimum program duration.

Since we now have a continuing need to support the space station, we cannot simply halt all manned space operations for six years like we did after the Apollo-Soyuz mission, in order to pay for new vehicle development. If NASA decides to continue use of the Shuttle, in spite of the return-to-flight schedule slippage and its steadily-increasing costs, they may be unable to afford adequate funding to develop both the projected Crew Exploration Vehicle (CEV) and a new booster. Development of a new, affordable booster by private companies may then be the only solution.

What kind of boosters and vehicles would the new program need for both Earth orbit and lunar operations? You do need a booster large enough to place complete spacecraft or spacecraft modules in orbit. For any spacecraft with a specified crew or cargo capacity, there is some optimum (and minimum) module size and mass, below which the cost of and the extra mass needed for a complex on-orbit assembly becomes counter-productive. However, vehicles designed for use in a vacuum and that never need to re-enter are much easier to modularize. Needed vehicles can be reduced to several basic types: Crew transfer vehicles (CTV) for use between Earth's surface and Low Earth Orbit (LEO) and return to surface via re-entry. Should support 6-7 crew members. Medium cargo vehicles (MCV) for use between Earth's surface and LEO and able to return some cargo and waste to the surface. (these could simply be stripped but pressurized CTV's with no life support and automated controls). In an emergency, they could even serve as "no frills" crew rescue vehicles. Medium Booster to launch CTV and MCV to LEO (actual payload capacity 20-25 tons). Heavy Lift Vehicle to put Lunar spacecraft and Lunar equipment in LEO (no return needed) (actual capacity 50-120 tons). Trans-Lunar Vehicles (TLV) for Crew and/or Cargo and for use between LEO and lunar orbit or Earth-Moon Lagrange-point (L1 or L2) location(s), and return to LEO for re-use, possibly using aero-braking to reduce energy costs on the return trip. Cargo vehicles could operate automatically with no crew cabin needed, or with a pressurized container for "shirt-sleeve" only cargo. Lunar Ferry (LF) to move crew module and/or multiple sizes of cargo to the lunar surface and return to Lunar orbit or Lagrange Point Bases for re-use.

The medium booster for a CTV or MCV could be supplied in a number of forms, by a variety of means. If there was a near-term (in six years) need for one, an existing EELV could be used, but there is no such near term need based on the mission schedule that NASA has published. (The no-shuttle scenario described above would be one). Using EELV's to support Lunar operations after 2014 might be financially unsustainable, violating the stated goals of the program itself. In the review of medium and Heavy booster options in ***Space News*** (4/12/2004, page 15), Robert Sackheim of NASA Marshall makes the argument that developing a "fully reusable launcher" is untenable due to the low launch rate. He describes an (initial) launch rate of 6-10 EELV's per year. This would presumably support just a single lunar mission per year, and ignores a later probable increase in the number of missions. The argument ignores the efforts made by NASA just a few years ago to develop the fully-re-usable but ill-fated X-33, and also ignores the possibility of building a partly-re-usable booster.

One form of this could be a larger version of one of the re-usable launch concepts already partly developed, such as a recoverable first stage launcher based on the proven DC-X, which takes off and lands vertically. This avoids the need for cutting-edge technology development and makes it more of a straightforward engineering task. Another form would use a large supersonic launcher aircraft with an expendable second stage and a CTV or MCV as payload. Again, this could be based on existing technology. An open competition to all companies with launch guarantees and mandated sharing of subcontracts with the losers of the primary contract would assure the availability of such a vehicle. Still a third version would be a fly-back first stage design similar to the Starbooster concept, which retro-fits an existing, proven first stage rocket to fly back horizontally and land like an airplane after a normal, vertical launch. Such a design accepts the loss of some payload capacity due to the weight of the wings and other equipment as the price for recovering the whole first stage. Additional recoverable booster designs are being worked on, and at least one private booster of sufficient size, the Falcon V, is currently being designed by Space-X to be "man-rateable."

At least the first of the three options listed above could also be used for the development of an HLV, a true Heavy Lift Vehicle (capable of lifting 50-120 tons to orbit). The aircraft based launcher and aircraft-like booster flyback options would be difficult to build due to the very large size of the airframes needed. The minimum cost method to get some kind of a cheap HLV (at least for minimum development costs) might be to privatize the remaining shuttle elements (not including the orbiters) to create a private shuttle-C or equivalent.

To assess the minimum HLV launch mass to support Lunar operations, we first have to back up and get a list of everything that needs to be launched for the Lunar program (including items 5 and 6 above), and to also provide reasonably cheap and reliable access to Luna. We then need to identify the heaviest, bulkiest items that cannot easily be launched in sections and re-assembled in Earth orbit or on the Lunar Surface. Some primary examples of what needs to be launched would include:

Empty and un-fueled TLV. Empty and un-fueled LF. Habitation and utility Modules with life support equipment (similar to current space station modules) for in-space use. Habitation, utility and lab Modules with life support equipment (similar to current space station modules) for use on Lunar Surface. Heavy excavating equipment for covering habitation modules on the Lunar surface. Pilot-scale mineral extraction and processing equipment. Scientific equipment (Geology, Radio-astronomy, etc.). Power source equipment for the lunar night (small nuclear reactor). Pressurized long-distance Lunar Exploration Rover. Collapsible crane to off-load and move heavy objects from cargo landers.

There are several different proposed methods for establishing regular traffic to Luna, but many of them would come at typically outrageously high "NASA" price levels. Having a larger number of vehicle types obviously drives up the development costs for the missions. Some development will be needed, but ways need to be found to reduce it. The best way is to use modules. This would also reduce the required payload size of the HLV.

The baseline modular vehicle is the LEM, or Apollo's Lunar Excursion Module, later called just "Lunar Module" or LM. The first manned spacecraft ever flown without an aerodynamic exterior, it had two expendable stages or modules. The Descent stage carried both stages to the lunar surface from lunar orbit. It could transport the Ascent stage with wo astronauts and over 1000 lbs. of equipment to the lunar surface, where it was left. The Ascent stage used a single very highly reliable rocket engine to transport the two crew members and several hundred pounds of lunar samples back into lunar orbit, where it rendezvoused with the Apollo Command Module and then was discarded.

I do not think that building giant disposable LEM's is the right way to return to the Moon. However, re-usable vehicles based on the LEM are a different story, and there are also other ways to do it.

Let's look at the facts and problems of a Lunar program, point by point.

We have about a decade or more before any real chance of a manned lunar mission, so there is no urgent need to settle on a final design right away. Since we already have had experience building one lunar vehicle, there is no reason to require ten whole years to design and build another. (2) Lets assume we want to do the first lunar mission in ten years, (by the start of 2014), and that we need five years to build and test the new vehicles. That gives us five years to pick a basic design strategy and design the vehicles, so the designs should be finalized by 2009. If we get funds for lunar operations sooner, the schedule could be moved up by a few years. The biggest obstacle to **starting** any kind of a lunar program is cost, and the biggest obstacle to **continuing** a lunar program is also cost. It behooves us to make a maximum effort in reducing both design and construction costs AND operational and maintenance costs. NASA planners have justified the use of expendable spacecraft designs in the past by claiming that design costs are reduced somewhat and they can be built faster. If you use expendable spacecraft, you have a very high operational cost of replacing each vehicle used in lunar space, which includes the cost of building each new vehicle AND transporting it to Earth orbit and then lunar orbit. Giant cargo-carrying copies of the LEM, capable of landing an entire habitation module or heavy excavator on the Lunar Surface, all individually hand-crafted and hand assembled, are going to be very expensive to replace. Unless there is a re-fueling and assembly base in Earth Orbit, it would take a very, very large HLV to launch the fully fueled super-LEMs plus their cargo. If your funding for new vehicles gets cut off, you eventually will run out of both launchers and vehicles and then you are grounded. We cannot guarantee how many boosters and landers we would get to build before Congress closed the production line again like it did to the Saturn 5 and Apollo. Then we would either have another 30-year gap of no human activity beyond low Earth orbit, or we would have to pay all over again to design and build a set of new vehicles. When a reusable vehicle has been used once, it will probably fly the next time safely. This is not true for expendable vehicles; each time they fly, they fly for the very first time on their "maiden flight", when hidden mistakes that could destroy a vehicle are most likely to be discovered . The needs of an initial series of missions and those of later missions are rather similar: the need to transport crew, cargo and hydrogen or other fuels to the lunar surface, and crew and eventually oxygen back to lunar orbit.

Thus there is no reasonable excuse to delay building a re-usable vehicle for use with the "permanent" base, that could also be used with a "temporary" base. We have the tools, materials, knowledge, and time; so if re-usable vehicles are feasible at all, lets build it right the first time. It therefore makes sense to spend at least a couple extra years if needed planning to build reusable vehicles, rather than rushing another expendable design into production. This would be like the Hollywood producer who rushes in to start set building and production for a movie before the script is even written.

When it comes to designing an actual lunar transport system, a series of major questions arises. Where is the best trade-off point between servicing re-usable lunar vehicles in space, vs. building new ones for each trip? Can we build a lunar vehicle that is very easy to service between missions, and which can check out its own systems reliably without requiring too much expensive crew time? Would it be cheaper to bring the vehicle down to the surface as cargo to service once every five missions? Would it be better to build a vehicle with a re-entry shell, so that it could fly itself down to be serviced when needed? (This is why several people in the 1990's even suggested a Delta Clipper type vehicle, re-entry shell and all, for use as a lunar ferry!). Is a more practical lunar ferry one used only between lunar orbit and lunar surface, or between Low Earth Orbit and lunar surface? Should we use high energy but difficult to handle cryogenic oxygen and hydrogen propellants for the Lunar Ferry, or non-cryogenic ones, providing less energy.

Obviously, the one thing that we really do need in Earth orbit for efficient lunar operations, no matter which type of lunar vehicle we choose, is a re-fueling and cargo transfer facility. Without such a fuel depot, a Moon-bound vehicle would have to serve as its own propellant depot, waiting in orbit to be filled up by several successive flights of tanker vehicles from the ground, and losing quite a lot of propellant from boil-off while waiting! The depot probably does not need to be permanently manned; the crews of the tankers, cargo and crew vehicles should be able to handle the fuel transfer tasks, but it should include a provisioned habitat module as an emergency refuge.

Careful analysis and probably some international negotiations will have to go into deciding what orbit the depot should use. An equatorial orbit would get a maximum payload out of each launch from the surface, but few large launch facilities, other than that of France in Guyana, now exist near the equator on an east-facing coastline, to take full advantage of this orbit. The disadvantage with any higher orbital inclination, such as using the current station orbit, is the smaller cargo mass for the other HLV's required. Since the departure angle from LEO to the Moons orbit varies so widely, (because the Moon's orbit is aligned only five degrees from the plane of the ecliptic, not from that of the Earth's Equator), delta-V requirements would vary considerably depending on the time of year. The maximum plane change from an equatorial LEO would be about 28 degrees. If you tried to launch to the Moon from the existing space station orbit of 52 degrees, when both the ecliptic angle and lunar orbit were at the maximum opposite or worst angle, you would need to make a plane change of up to 80 degrees. This problem would make lunar missions virtually impossible for a large part of the year.

The Earth orbit depot should be heavily protected against impacts from space debris, and it should be able to use solar energy to re- liquefy all of the boil-off of the cryogenic liquid hydrogen and oxygen propellants. It must allow safe transfer of propellant from tankers to storage tanks and then to lunar ferries in a micro gravity environment. This includes a means of forcing propellant out of a tank when gravity is not forcing the liquid floating in the tank against the exit valve. One example of the way this was done for Apollo by Bell Aerospace (with non-cryogenic propellants) was the use of Positive Expulsion Tanks. Each tank contained a reinforced silicone rubber bladder, which expanded like a balloon against the inside tank walls as fuel was pumped into the bladder. To force fuel back out in zero gravity with no pockets of gas within the fuel, nitrogen gas was forced under pressure between the tank wall and against the other side of the liner, squeezing the propellant out like toothpaste! Cryogenic propellants are probably too cold to use with the flexible Positive Expulsion bladders, but there are other methods. In-orbit fuel transfer is one area where much more work needs to be done.

Using a Delta Clipper (SSTO) type vehicle for a lunar ferry would pose some unique problems. The vehicle would be launched into orbit without cargo, but would be able to carry a significant amount of cargo on lunar missions. The propellants and cargo would need to be loaded in orbit. The main advantages are (1) no parts or stages are expended, (2) only 1 vehicle type is needed), and (3) the vehicle CAN be serviced on the ground, instead of in space, when needed, by simply flying it down to the ground. (It could use aero-braking when returning from the Moon to reduce the final re-entry speed and heating. It can also put itself back in orbit.) Some disadvantages are: (1). We would need to develop a real SSTO (in addition to other boosters) that can at least put itself (without any cargo) into orbit, (2) The SSTO ferry would have a lower lunar payload to weight and fuel ratio, compared to the standard dual system using (a) lunar transfer vehicles (to reach lunar orbit), and (b) lunar ferries (to reach the lunar surface). (3) Compared to a non-aerodynamic (vacuum-only) vehicle, the size and shape of cargo payloads carried by an SSTO ferry would be greatly restricted by the size and shape of the re-entry shell, so carrying bulky habitat modules would be very difficult.

There are a number of more standard strategies in designing a set of lunar vehicles. One uses the vertically modular lunar ferry, based on a re-usable LEM, in combination with a LTV using the same exact design for the crew cabin. The LTV propulsion module itself could also be very similar to that for the descent vehicle, minus the landing equipment. Remove the crew cabin module from either an LTV or LF crew vehicle and you have a cargo vehicle.

The basic reason for using modularity is to reduce the number of different types of vehicles you need to design. There are two types of modularity discussed here, horizontal and vertical. The LEM has incomplete vertical modularity, but we could have built an unmanned cargo lander out of the LEM descent stage. This could have delivered a habitat to the lunar surface, and allowed astronauts an extended stay on the surface in the early 70's. The new lunar ferry would seem to have a very similar configuration, but in reality would operate differently. Only one module type would carry engines, and no modules would be discarded. It could have a passenger or cargo capability by designing a lander module (containing the engines and landing gear) that can handle a centrally mounted passenger cabin module, deliver a large centrally mounted lunar habitat or deliver two-side mounted cargo containers. (For a vehicle that lands vertically, it is very important to balance the load). For cargo, it will also be important to be able to easily off-load cargo, including habitats and machinery too large to go into a container, so an early cargo item at each landing site would be a collapsible crane. Unlike the LEM, the ferry would use the same engine both to land and take off for a return to lunar orbit.

Computer hardware and software would be an integral part of such a system. Each independent module would be able to land and return by itself. In addition, software and hardware sensors, built-in test equipment and alarms would have to take a much larger role in assisting the crew in providing maintenance and verifying that the vehicles are still safe to fly. Depending on the location and cost, tiny video cameras, scanning and other types of test equipment might also be built-in place, to allow the crew to inspect areas difficult or impossible to access when in space.

Another aspect of modularity on the functional scale is equipment modularity. If you have several vehicles, and a critical piece of life-support or other equipment is damaged, you would be able to use an identical unit from a vehicle, which had exceeded its flight lifetime, and had been left at a lunar base or in orbit just for spares. Re-usable spacecraft would not be re-used indefinitely, except in an emergency and would have an established flight "lifetime". When not being used, the spacecraft could be stored at a depot or on the surface. They would be protected by easily removable thermal and anti-meteorite/space debris "blankets". It might be possible to design a curved telescoping "blanket" which would surround a docked vehicle automatically.

The use of in-situ or local resources to support operations should be a major development priority. For that reason, small expendable unmanned landers should be used to deliver instruments, habitats and/or prototype mining equipment to the lunar surface before the first new manned expedition, which I would hope would be international in scope and support. With only a 2.6 second round trip delay between Earth and Luna, some of this test equipment, such as miniature loaders to load samples into the test processing units, could be tele-operated from Earth. By the time lunar manned missions resume, actual tests of the equipment on the Moon could allow us to build a large working prototype of an oxygen smelter ready for use by astronauts.

Whatever its design, the initial ferry must be able to carry its own takeoff propellants down to the surface, since there will be no lunar oxygen plant there initially. Here is another place where modularity comes in. The very first flights could use extra modules to carry the takeoff propellants. Then when a lunar oxygen plant is operational, fewer modules would be needed for the same cargo or mission. For early missions with the more orthodox LEM-like ferries, using them in pairs would also allow a larger payload to be landed. One ferry would descend to the surface with a cargo of propellant, the other with a large payload, such as a habitat module. The second ferry would then be re-fueled from the first, and both would return to lunar orbit. With this method, a ferry could be sized just large enough to land with the maximum expected cargo weight. For most early missions, each ferry would carry enough propellants to reach orbit without re-fueling, and would carry a smaller payload. Rescue capability could be provided by having more than one vehicle available at each location where crews visited or worked. If a vehicle became stranded in orbit, it would be relatively simple for a single pilot to use another vehicle to rescue them.

By the time the oxygen plant is running, there will probably be a "symmetrical" propellant supply system in place. Lunar transfer vehicles will bring lightweight hydrogen fuel to a lunar orbit or Earth orbit depot from Earth. Lunar ferries will be loaded with the hydrogen. They will bring extra Earth hydrogen down with them when they land, to use when they take off and will then be loaded with extra lunar oxygen when they are ready to take off. After they reach orbit, they will be able to use the extra oxygen to retrofire and land the next time, so that no heavy and expensive oxygen (8 times heavier than hydrogen), will need to be carried up out of Earth's steep gravity well. In this situation, the oxygen tanks are only full just before liftoff from the Lunar surface, and the hydrogen tanks are only full just before leaving Earth or lunar orbit for the Lunar surface. In this symmetrical system, all the Hydrogen comes from the Earth and all the oxygen comes from the Moon. For safety, the cryogenic hydrogen and oxygen tanks should be well separated from each other.

There are arguments against using a Lunar Orbit Fueling Terminal, or LOFT, similar to the ones used against an Earth orbit depot. The first argument is that it might have to be continuously manned, and the crew, being outside the Earth's protective magnetosphere, would get a high dose of solar radiation, thus the facility is a bad idea. A good rebuttal is that, just like the LEO equatorial depot, the LOFT needs to be tended only by visiting crews, who can handle the fuel transfer themselves. However, there would still be a need for a heavily shielded emergency habitat or refuge at a lunar orbit depot, since a solar storm could hit before a crew had time to reach a lunar surface base.

The second argument against a LOFT is that many missions will be exploratory ones to higher lunar latitudes; areas that cannot be easily reached from an equatorial orbit, and only during part of the month from a polar orbit. The rebuttal is that since there are no concentrated or localized ore bodies as such on the Moon (that we are aware of), there is every reason that an initial lunar mining base should be either close to or right along the lunar equator or close to or at one of the lunar poles. In these locations, a transfer station and Earth return vessel in a matching equatorial or polar orbit would pass overhead, available for a trip to the surface on every orbit. This also means a lunar ferry doesn't always need enough fuel to reach any point on the lunar surface, but just the points accessible from either equatorial or polar orbits. Most ferry missions will not be exploratory, they will be support and crew transfer missions to just a few base sites. Exploratory missions would need to carry extra fuel and less cargo to reach other areas.

The last major argument against a LOFT is the same one directed against using the Space Station: fuel transfer will be too time-consuming, expensive and difficult. The rebuttal here is the same as before: the facility will be man-tended only, there will be no scientists worrying if their experiment is being joggled, and the facility could be also be built in a modular fashion, so that failed storage and transfer units could simply be replaced if repairing them in space is too difficult and time-consuming.

If we are extremely lucky and do find usable ice deposits at the lunar poles, we would have access to a ready supply of oxygen and hydrogen. Current expectations are that there are no solid ice deposits, but that there are a few percent of loose ice crystals in the soil of the shadowed areas, which is probably enough to extract and use locally. (This does not mean that extracting the water will be easy.) It now seems unlikely that there will be enough water found for a polar base to be able to supply Earth orbit operations with propellants. There are also other non-rocket alternatives for lunar cargo handling. A lunar mass driver could easily deliver "smart" canisters of propellant, ore, or "whatever" into lunar orbit or the L2 point, where they would queue up to discharge their payload, and be returned to the surface in batches via ferries, ready for another trip.

Once decisions have been made on the design of a set of lunar vehicles, the minimum payload requirements for the HLV can finally be determined. If a SSTO based lunar vehicle is used, the HLV's would still have to be large enough to carry habitat modules and other heavy objects into orbit, and the SSTO would have to be large enough to land it on the Moon. An HLV used with this option would not need to carry any lunar spacecraft into orbit, but it would need to be able to carry fuel and cargo to the orbital depot. If modular Ferry and LTV designs are used, the HLV must have enough capacity to lift the largest and heaviest un-fueled, empty module into orbit, as well as cargo and fuel.

The remaining question then becomes one of how to create a true but affordable and reusable HLV to lift the largest objects into LEO. The primary issues are determining the operational costs and reliability of such a vehicle compared to the existing shuttle system and also compared to a new privately built HLV. The costs should include development, construction and operational costs over a minimum period (established before) of 25 years from the start of HLV launch operations). It has been estimated that a shuttle-based HLV would cost about $4.5 billion to design, (a little more than a year of shuttle operations.) Such a vehicle, depending on whether the main engines are recoverable or expendable, the number of solid rocket segments used, and the total weight and design of the shroud, would be able to lift between 70 – 110 tons. This is three to four and a half times larger than the Shuttle's maximum payload of about 25 tons. It is possible that a private design might be able to do better than this, depending on the cost per launch. I have not seen good figures for a single Shuttle-C launch, or the annual cost of maintaining the staff to operate the launch system. Mike Griffin, our new NASA Administrator, who suggests use of a Shuttle-C, rather than EELV's in his ***Space News*** article of 4/8/2004 (p. 13), gives a "recurring cost of about $1 billion, but it is not clear what kind of cost he refers to. This critical piece of information needs to be determined by several different groups of engineers and managers working independently. Removal of the need to maintain the complex Shuttle orbiters themselves should significantly reduce the annual cost, but is this reduction enough? If the annual operations cost remained over 2.5 – 3 billion a year (amounting to $25- $30 billion per decade, $62-$75 Billion per 25 years)—irrespective of launch rate, a new privately built HLV would make more sense. True privatization of any Shuttle-C system would be mandatory to guarantee the predicted costs under contract.

Relatively cheap, safe and reliable access to Luna is critical to any reasonable scheme for development of the Moon and Cis-lunar space, and even though we may differ in our approaches to it, we should all work to see that such access is made available during the next decade.

John K. Strickland, Jr. joined the American Rocket Society as a student member in 1961. In 1975-6 he became a member of the National Space Institute and the L-5 Society, "parents" of the NSS. He serves as chairman for the Austin Space Frontier Society from its founding in 1981 to the present. In 1988 he was creator and designer and since as the coordinator (and sometimes builder) for the Heinlein award, and worked on the design and production of the Von Braun Award. In 1988, Strickland was a founder of the NSS Chapters Assembly, and have served as one of its officers. He is a director of the Sunsat Energy Council, and an active member of several other pro-space organizations. His interests include access to space and re-usable spacecraft, space policy, space solar power, and planetary and lunar base infrastructure. Strickland has written many articles including a chapter in the 1998 edition of Dr. Peter Glaser's book on Solar Power Satellites, and in 2003, a chapter for the new Boy Scout Merit Badge Manual on Space Exploration. He has been a Delegate to the Texas State Republican Conventions (2000 and 2002), and worked to insert specific pro-space wording in their party planks.

The following essay shows why the "magnificent desolation" of the Moon may turn out to be one of its best assets. Pete Worden recently retired as a General from the Air Force, and his thinking often reflects the military mind. During his tenure there he was able to drive two major pro-frontier projects through to fruition, the DC-X vertical take-off and landing vehicle, and the Clementine Lunar Orbiter. Both of these had a significant impact on the space settlement movement, and the Clementine mission was able to spot what appeared to be ice on the Moon's poles—a huge and important finding. In both cases, Pete was leveraging off of national security needs to help open the frontier. Although it may reflect a bit of the "dark side" of human endeavours, such activities as he outlines here are the almost inevitable outcome of our march forward into the future...(although as the band Pink Floyd says: "there is no Dark Side of the Moon...")

Lunar High-Risk Manufacture

by Simon P. Worden

INTRODUCTION

In a few decades, the technology to replicate the human brain, or at least construct an analog computer system with similar levels of complexity, will mature. This may be exciting to some. But to others, the emergence of something resembling artificial intelligence raises visions of evil machines highlighted in dozens of science fiction thrillers. If mankind were to develop true artificial intelligence (A.I.) it would revolutionize our lives. Some argue that a hybrid existence where the human brain is augmented and extended by artificial intelligence could enable an unlimited existence in space and time—perhaps even defeating death. But those who fear the rise of aggressive entities such as Star Trek "*Borg,*" bent on absorbing all independent entities object to this future. How is this dilemma related to space exploration or the Moon in particular?

On January 14, 2001, President George W. Bush announced a renewed commitment to extending human presence into the solar system, beginning with the Moon and moving on the Mars. This initiative has raised little enthusiasm outside traditional aerospace industry and space enthusiasts. NASA, as any government entity is gearing up to fight the last war—in NASA's case a repeat of Apollo. It is preparing grandiose plans for a second Apollo program featuring new space launch vehicles and manned capsules for journeying to the Moon and Mars. These plans have excited little interest in the U.S. Congress or in the public. The large cost of manned exploration—some critics put it at one trillion dollars or more over the next few decades—further decreases interest. In the case of the Moon critics further cite a "been there done that, got the T-shirt" argument. NASA says that the Moon is really a testing ground for advanced Mars exploration technology. But the public seems little interested in Mars either. Moreover the technology to mount a human mission to Mars is immature.

A new factor in human space exploration is the emergence of a true space private sector. It's important to understand the difference between "commercial" and "private" space investment. Current commercial space firms, such as Boeing, Lockheed-Martin and Northrop-Grumman hardly fit the description of private sector. These enterprises are similar in many ways to Soviet-style design bureaus. With a revolving door between government officials and company officers these firms are in reality a part of the government. There are, however a number of new companies in the space business as true entrepreneurs. Such companies are investing in, and making money from, new space goods and services such as imaging from space, navigation services and communications. But the most interesting new aspect in space exploration is neither the government design bureaus nor the new space companies. The new factor is a group which one might call "venture philanthropists." These are individuals, often billionaires, who made lots of money in non-space related high technology such as the Internet but retain a strong interest in space. Many of these people are relatively young. More to the point they grew up watching Star Trek and Star Wars and want to make the dreams real. And they have the money to do it!

The recent 2004 Ansari X-Prize winning flights of Burt Rutan was financed by one of these venture philanthropists—one of America's wealthiest men, Paul Allen, co-founder of Microsoft. But these philanthropists are not in the space business just to do neat things. They also have in mind making money—perhaps not on short term time scales, but eventually on huge volume scales. The emerging field of space tourism is their initial goal. The joint venture between Paul Allen and Burt Rutan's Mojave Aerospace Ventures, builder of award-winning SpaceShipOne and Richard Branson's (of Virgin Airways fame) Virgin Galactic Corporation intends to make lots of money off space tourism.

Some futurists envision private sector development of the Moon, perhaps for space tourism. But is this realistic? If a suborbital flight costs several hundred thousand dollars it's hard to imagine a lunar flight costing less than hundreds of millions—even with true private sector ingenuity. Space Adventures Corporation is marketing a lunar fly by based on low-cost Russian hardware. But this trip will cost over $50M. The customer population for this sort of tourism is clearly small. Others think the Moon could provide resources such as Helium-3 to power fusion reactors. Since working fusion reactors are at least 20 years off—as they have been for the past 50 years, it doesn't seem likely that this market will generate much private interest.

There is, however, a potentially lucrative use of the Moon, as a location for high-risk technology development and manufacturing which can't be done on Earth. The development of technologies such as the cyber intelligence mentioned earlier just won't be allowed on Earth. Yet the payoff of successfully developing and perfecting these dangerous technologies could be enormous. The Moon, using advanced robotic technologies, would be the ideal site for high-risk, high payoff development and manufacturing.

DANGEROUS STUFF

Along with cyber intelligence there are a number of other technologies that increasingly alarm some segments of the public. Leading this list is nuclear technology—the traditional bogeyman. However, other scary science threats are rising in public consciousness. Foremost among these new technologies are bio-engineering. Across the world groups are increasingly opposed to genetic engineering. Many nations have or may soon ban genetically engineered plants—even those that can increase food yields many times over to forestall famine. Particularly severe are critics of manipulations of the human genome. Perhaps the ultimate concern has not yet been highlighted much, although it's been thoroughly vetted in science fiction. This is the simultaneous development of artificial intelligence with nanotechnology. The combination of these technologies raises the specter of intelligently driven, self-replicating "nanobots" as the ultimate plague.

We must balance against the bad scenarios the immense possibilities they offer for progress. For example, nuclear technologies are possibly the best bet for solving the energy crisis. Like it or not we will soon be out of oil. Alternate fossil fuels may postpone the crisis a number of decades—but at a serious, perhaps even fatal cost to the environment. Some warn that continued use of fossil fuels through the 21st century could trigger irreversible global changes plunging the planet into a heat-induced hell similar to Venus' surface or into a completely frozen state that the Earth apparently endured early in its evolution. No new nuclear plants have been built in the United States for decades—and few anywhere else. The problem is the lack of safe, modular and affordable designs. But even the research for such new nuclear technologies, including nuclear fusion is stalled, often for environmental and safety concerns.

Bio engineering is another area of great promise. Not only can bioengineering feed the world, but it promises cures for the most serious death-dealing illnesses. Yet increasingly stringent, some would say prohibitively restrictive procedures are being placed on bioengineering. Nowhere is the concern more emotional than that surrounding the human genome. One example relevant to space colonization is living on planets such as Mars. Many talk about modifying Mars to support Earth life. Yet an alternative is to alter terrestrial life to better survive on Mars. One problem is the possibly unavoidable degradation of human physiology caused by low gravity. We do not yet know if Mars 1/3 gravity will cause animal physiology to degrade as zero gravity does. One solution would be to alter human and animal genomes to counteract this effect. Current angst over the release of "Franken" life on Earth may permanently relegate such possibilities to science fiction.

Nowhere is the potential for new technology as two-edged as the intersection of nanotechnology and cyber technology. An excellent review of the issues is in a 2002 book entitled, Ray Kurzweil vs. the Critics of Strong A.I., edited by Jay Richards. The argument that human-level complexities in artificial computers can emerge in a few decades is controversial. Even more interesting, and distressing to some is the possibility than micro robots could be injected into and read out the entire contents of a human brain. This information might subsequently be installed in the human level-of-complexity supercomputers of the future. Aside from the metaphysical issues of whether the resulting entity is conscious or "human," such a direction, if feasible, might lead to unlimited lifetimes for all. This Godlike future doesn't excite all experts. Sun Microsystems co-founder and Chief Scientist Bill Joy wrote an essay in the April 2000 issue of Wired Magazine entitled "Why the Future Doesn't Need Us," opposing future research on the proposals for artificial intelligence outlined above. Joy postulates humans being replaced by machines or, perhaps worse a laboratory experiment involving self-replicating nanobots becoming all-consuming "grey goo" snuffing life on Earth. He and many others suggest foregoing these new technologies for an indefinite time.

I, and I suspect many readers of this book, find the appeal of a greatly prolonged existence compelling, despite the risks. Each new birthday and ache and pain underscores mortality—a technological solution seems to warrant some attention and risk. The question is how much risk is acceptable? I propose we export development of these dangerous technologies to the lunar surface. We could establish a technology development reservation on the Moon with significant controls to assure escaped products or faulty results don't back-contaminate Earth. While no quarantine is perfect, the lunar surface is a pretty good isolation ward. This approach might enable the development and testing of incredible new technologies while providing critical margins of safety—not only in space, but also in time. In the more distant future, as technologies further develop, the reservation may also be moved out further, to Mars and even beyond.

THE LUNAR TECHNOLOGY RESERVATION

The Moon is an ideal location for dangerous technology development. It lacks hydrologic and atmospheric transport mechanisms so mistakes stay where they occur. The naturally harsh radiation, thermal and vacuum environment is an outstanding sterilization medium for biological agents. The gravity well would keep problems from migrating easily back to Earth. Finally, pictures of the Moon look remarkably like a nuclear war already occurred so the usual concern over green issues would likely be less, but undoubtedly not zero.

A proposed way ahead is to establish an industrial reservation at the lunar south pole. While the Outer Space Treaty prohibits nations, such as the United States, from claiming or owning land on the Moon or other solar system bodies, it arguably permits private concerns to do so. Proceeding from this premise, what should the U.S. Government's role entail, or any other national or international organizational role?

There are three things a government could do to promote private lunar development: first, it could establish the necessary infrastructure. Second, it should develop general technologies. And third the government should establish a legal, security and safety framework for lunar development.

The first step is to put in place the necessary infrastructure to support lunar exploration—and more important exploitation. This program differs from that originally envisaged by NASA following the President's space exploration initiative in that it focuses on the necessary infrastructure for further exploration such as communications, navigation and resource mapping rather than government human exploration programs and science. Fortunately, the technology and capability exists to ensure early effective supporting capabilities for the Moon. Over the past decade the technology of "microsatellites" or "microsats" has matured. Using the same miniaturization we've seen in the computer industry the capabilities of a satellite which weighed thousands of kilograms a decade ago may be done with microsats weighing several hundred kilograms today. In addition, these microsats can be built for modest money—a few tens of millions of U.S. dollars per satellite vice hundreds of millions for larger systems. In addition, there are a variety of low-cost launchers available to put these satellites in place. One option—used by the Europeans extensively is to piggy-back microsats as secondary payloads when we launch a large satellite. The U.S. is now modifying the new EELV (Evolved Expendable Launch Vehicles) developed by the U.S. DoD for just this purpose. Launch in this case is cheap as the primary payload bears most of the cost. Alternatively, new small launch vehicles now under development by the DoD and several private sector groups could launch these systems at low cost.

The international community is reeling somewhat over the U.S. announcement of the President's new exploration vision. They fear that the United States will abandon international cooperation and the International Space Station—leaving them holding the bag. A properly constructed U.S. led effort to put the necessary utilities in lunar orbit could counter this perception and generate considerable international enthusiasm. A number of nations including the European Community, Japan, China and India have or soon will launch robotic lunar missions. Cooperating in putting the necessary infrastructure in place to support further lunar use and exploitation could be an appealing global goal. Many would also view it as an opportunity for the world to re-invent how to provide terrestrial services such as global positioning, without the constraints of previous infrastructure.

A 2004 White Paper by U.S. astronaut Lee Morin suggests how the government might develop technology. Dr. Morin believes that lunar exploration should begin immediately by developing and deploying robots on the lunar surface. These robots could be delivered via existing launch vehicles and need not wait for new crewed or heavy lift launch capabilities. Current U.S. EELV's can place 1000kg or more on the lunar surface. This is sufficient to begin a robotic infrastructure focused on developing and using in-situ resources, including the polar water and other volatiles.

A robotic approach has the goal of increased lunar robotic capability. Ultimately we'd like to have robots on the Moon capable of manufacturing and maintaining more robots. As the infrastructure grows, less and less components would be needed from Earth. Initially we might focus on manufacturing 19th century technology components such as basic robot housing and simply manufactured structural components. As more and more skillful robots are produce the robot colony would evolve to higher technology manufacturing. For quite some time sophisticated 21st century technology such as central electronic components would be provided from Earth. Tele-presence using Earth-based human operators would also play a major role at first. Over time the percentage of robotic construction and activity relying on support from Earth would fall. Eventually, virtually nothing would be needed from Earth. Once in place this robot colony could form the basis for the dangerous technology reservation discussed above.

The third and most important role of the government would be protection. First the government would need to protect the rights of individuals from legal encroachment. This could best be done by a major power—initially probably the United States asserting and protecting the rights of private usage of the Moon. But the longer range role of the government probably means physical quarantine. It would be best to have governments inspect technology development to determine when things are going awry. Further we'd like governments to have the means of terminating an experiment if necessary. This may, indeed, be the first true need for military force in space.

As a footnote it's important to understand that some technologies, particularly those of an A.I. nature could evolve quite quickly. Machine intelligence is very likely to operate millions of times faster than a human brain. While a lunar dangerous technology reservation provides significant quarantine in space and time, nothing is perfect. From the start mechanisms to monitor and terminate untoward development must be in place. Excessive dawdling in decision making must be avoided.

SUMMARY

Unless and until a significant product is found from space it's unlikely that space expansion will occur. Currently human space exploration is preceding in a manner similar to past human space endeavors—as exercises in national and international political resolve. This problem is recognized. The United States and other governments assert that new exploration will focus on private sector development. But it's difficult to see how this will occur unless the private sector sees real profit from space in products and services on the ground.

The development of dangerous technologies on the Moon could be just the ticket for opening space to the private sector. Dangerous and perceived dangerous technologies with immense promise such as nuclear, biological, nano- and cyber technologies offer great results and profit. The Moon, particularly its south pole with its ready source of continuous solar power and resources including water and other volatiles, is an ideal place for this dangerous technology reservation. The important roles of governments would include developing the necessary communications, navigation and mapping support and technology development, particularly in robotic self-replication. However, the most important role of government in the private development of the Moon is to protect the entrepreneurs from government and protect the Earth from mistakes the entrepreneurs make.

Space offers an unlimited future. This has always been a claim of space zealots. But in recent years newer glamour technologies have taken center stage with promises of nearly eternal life. Little wonder that space technology is increasingly seen as an outmoded steam engine level effort. But the beautiful rose of the these newer non-space technologies clearly has thorns—probably very poisonous thorns. Using the Moon as a location for removing the thorns may be the key to all our futures.

Brig. Gen. Simon P. Worden (retired) is currently serving as a Congressional Fellow with the Office of Senator Sam Brownback (R-KS). He is responsible for supporting the Senator on space and NASA issues. He was appointed as a Research Professor of Astronomy at the University of Arizona on 1 May 2004. General Worden retired in 2004 after 29 years of active service in the United States Air Force. His final position was Director of Development and Transformation, Space and Missile Systems Center, Air Force Space Command, Los Angeles Air Force Base, Calif. He was commissioned in 1971 after receiving a B.S. from the Univ. of Michigan. He entered the Air Force in 1975 after graduating from the University of Arizona with a doctorate in astronomy. Throughout the 1980s and early 1990s, General Worden served in every phase of development, international negotiations and implementation of the Strategic Defense Initiative, a primary component in ending the Cold War. He twice served in the Executive Office of the President. As the staff officer for initiatives in the George Bush administration's National Space Council, the general spearheaded efforts to revitalize U.S. civil space exploration and earth monitoring programs. General Worden commanded the 50th Space Wing that is responsible for more than 60 Department of Defense satellites and more than 6,000 people at 23 worldwide locations. He then served as Deputy Director for Requirements at Headquarters Air Force Space Command, as well as the Deputy Director for Command and Control with the Office of the Deputy Chief of Staff for Air and Space Operations at Air Force headquarters. Prior to assuming his current position, General Worden was responsible for policy and direction of five mission areas: force enhancement, space support, space control, force application and computer network defense. General Worden has written or co-written more than 150 scientific technical papers in astrophysics, space sciences and strategic studies. He was a scientific co-investigator for two NASA space science missions.

Dr. Patrick Collins was one of the first people to make the point that "tourism" might actually be a major economic driver in our quest to open the space frontier. Tourism is a multi-billion dollar industry on Earth, funding entire economies, i.e., Las Vegas, Hawaii. Yet the idea that there might be a market of regular people willing and able to buy tickets into space was long derided and ignored by the traditional aerospace establishment. That is, until just a few years ago, when inspired by his work and that of many other "believers," I signed up Dennis Tito to fly into space. Now, with the flight of Spaceship One and tickets already being sold for sub-orbital flights into space, we see the beginnings of a real industry, based on selling the "once in a lifetime" adventure of entering a domain once restricted to astronauts and cosmonauts. In this essay, Dr. Collins takes a not so giant leap, projecting that once our appetites are whetted, we will want to keep going higher and farther—including to the Moon.

The Future of Lunar Tourism

by Patrick Collins

Travel to and from the lunar surface has been known to be feasible since it was first achieved 34 years ago. Since then, there has been enormous progress in related engineering fields, so there are no fundamental "technical" problems facing the development of lunar tourism—only investment and business problems. The outstanding near-term problem is to reduce the cost of launch to low Earth orbit, which has been famously described as "halfway to anywhere." Recently there has been progress toward overturning the myth that launch costs are high because of inescapable physical limits, as companies are planning sub-orbital flights at 0.1% of the cost of astronaut Alan Shepard's similar flight in 1961. Market research shows strong demand for both sub-orbital flights and orbital services, and travel to the Moon will offer further unique attractions. Far from being a trivial topic which "real" space engineers should ignore, it is the key to making space and lunar development profitable—and so unstoppable. Allow me to use the example of Hawaii, where this paper was first presented at a Lunar Conference. Tourism is the largest business activity in the state, generating huge wealth not only in Hawaii, but also on the US mainland and in other places from where people trade or invest in Hawaii. All this wealth creation starts for the simplest, most human of reasons: People enjoy being there. Many millions of people have found that spending a few days in Hawaii makes them feel good. At first, people visited Hawaii spontaneously for its delightful climate and scenery; this inspired entrepreneurs to work to make it convenient and affordable for more and more people to visit. This has involved using their ingenuity to supply an ever-growing range of popular services, and has included supporting local governments to enforce regulations as needed to protect the environment that visitors want to experience.

Lunar tourism will be the same: as soon as they can, many people will travel to the Moon for the same reason—they will enjoy visiting there. Since the idea of space tourism is known to be very popular; since the Moon has a unique place in the mythology and traditions of every culture; and since there will clearly be many unique experiences during a trip to the Moon and back, it's clear that it has the potential to become a major tourist destination.

Unfortunately, many people in the space industry suffer from the mistaken idea that tourism has no economic value. They believe that, unless people are working to make some kind of machine, their work is not really valuable. This belief is objectively wrong; the error of the "labor theory of value" is a long-standing issue in economics: work to supply a product is not valuable if there is insufficient demand. To give a simple example, without demand for tourism services from billions of people handled by hundreds of airlines operating thousands of airliners, aircraft manufactures could not produce them at a profit, thereby together creating millions of jobs in the civil aerospace industry. By contrast, making machines which no-one wants to buy, however technologically advanced they are, actually destroys wealth instead of creating it, because it wastes skilled humans' efforts.

The wealth in Hawaii generated by tourism depends on people continuing to want to visit. And that can fall for a number of reasons—for example, if there is a war, or a recession, or if the local government allowed the environment to be polluted, or if businesses there fell behind other tourist destinations. But demand in any industry is vulnerable to disruption and competition—as the rapid shrinking of US manufacturing employment, including particularly aerospace, shows clearly.

Because of this way of thinking in the space industry, many of the general public have a "taboo" about the subject of lunar tourism, and even orbital and sub-orbital tourism. They find it hard to imagine travel to and from the Moon becoming an important part of the travel industry. They consider the idea futuristic—"maybe 100 years from now"—forgetting that it was already done more than 34 years ago. So let's clear up some "myths" about space tourism.

MYTH 1: "LUNAR TOURISM IS IMPOSSIBLE"

First of all, it is certain that travel to and from the Moon is possible—because it *was* done 34 years ago. It is quite hard to list all the products that did not yet exist in 1969—not just recent inventions like CDs, laptop computers, the internet, mobile telephones or carbon nanotubes, of course, but back in 1969 Boeing 747s, optical fibers, video-cassettes, the Walkman and even electronic calculators were yet to come; most people had never even seen a color television. Since 1969, there have been literally generations of the fastest technological progress in history—in materials engineering, production engineering, combustion engineering, semiconductor technology, computing, communications and many other fields. So anything that was possible 34 years ago is potentially very much easier today.

Since 1969 there has been about $1 billion of research in lunar science and engineering, well summarised in the collected proceedings of the ASCE's unique series of conferences at Albuquerque. Technically there are no fundamental unknowns about lunar development—except how rapidly the travel market will develop, and how cheap lunar travel may ultimately become as passenger traffic builds up to large scale.

MYTH 2: "IF SPACE TOURISM WAS POSSIBLE, SPACE AGENCIES WOULD HAVE ALREADY DEVELOPED IT"

Many people today believe that the fact that space agencies have not developed passenger launch vehicles proves that they are impossible to produce with current known technology. This is perhaps the most damaging myth, but it is not true, as shown by two recent events.

First, Dennis Tito demonstrated in April 2001 that, despite having spent $1 trillion of taxpayers' money over the past 40 years, the space agencies of the USA, Europe and Japan have not reduced the cost of getting to space by a single cent. It is hard to believe that any American learning about Gagarin orbiting overhead in 1961 would ever imagine their government would spend nearly $1 trillion over 40 years—and yet the cheapest and safest way to get to space would still be to use the same rocket that carried Gagarin! The Soyuz rocket has been unmatched for 42 years—and surely will be for more than 50 years; this may be unique in the history of transportation. It's also very important to understand what this astounding fact means: among other things, it proves that space agencies have not been trying to reduce the cost of access to space. The truth is that they have no interest in doing so.

More recently, thanks to the efforts of X-Prize winner Burt Rutan and his competitors, it is being demonstrated that sub-orbital space flights in the near future will cost 1/1,000 of what it cost to launch Alan Shepard to space in 1961. That is, by applying some of the engineering knowledge that has accumulated since 1961 we can reduce sub-orbital transportation costs by some 99.9%.

Moreover, sub-orbital passenger flights using reusable vehicles could have started during the Apollo project, since the technology needed for this was developed during the 1960s using rocket-planes. Thus, if space agencies had economic rather than political objectives, a commercial sub-orbital space line could have been in operation before humans stepped on the Moon. But this did not fit space agencies' thinking. Instead, they focused on performing government space "missions." As a result, in 2005, most people, including politicians, journalists and scientists, still mistakenly believe that traveling to space is almost impossibly difficult and dangerous, and so expensive that it can be justified only for specially selected government employees. Once this is shown to be wrong in the case of sub-orbital space travel, it is to be hoped that the public will quickly understand that it is wrong also about travel to orbit and beyond.

The available evidence suggests that orbital flights, too, are amenable to similar cost-reduction through the use of some of the technology that has been developed during the 42 years since Gagarin's flight. The privately supported Space Tourism Study Program performed between 1993 and 2002 by the Japanese Rocket Society concluded that the "Kankoh-maru" SSTO VTOL passenger vehicle could be developed and put into service within 10 years for about $10 billion . This result is broadly endorsed by other experts on VTOL systems, including Dietrich Koelle and William Gaubatz, and it is broadly similar to cost-estimates for two-stage HTOL vehicles by groups such as Bristol Spaceplanes. It is noteworthy that neither of these vehicle types were considered in NASA's Highly Reusable Space Transportation research project, since they were judged not to meet that study's requirement to perform 100 flights/year carrying a 10 ton payload to orbit at less than $400/kg. However, that study did not consider producing a fleet of tens of vehicles and operating them to make thousands of passenger flights/year.

A former head of NASA's Office of Advanced Space Projects, Ivan Bekey, summarized the likely development costs of orbital passenger vehicles. For about the cost of developing a new airline, some $10 billion, i.e., far less than a single year of OECD space agencies' budgets—it is possible to produce a passenger launch vehicle which, when put into series production, could carry passengers to orbit for less than $50,000. In civil aviation, the cost per passenger of a mature air transport system is about three times the cost of the propellants; this rule of thumb will also apply to passenger space transport systems when they reach maturity. Importantly, the vehicles used for such services will become the basis of lunar tourism: since the velocity-change for a return-trip between LEO and the lunar surface is less than that between here and LEO, a VTOL launch vehicle, if refueled in LEO, could make a return flight to the Moon's surface—though in practice it would be suitably modified.

In view of these considerations, it is necessary to face the fact that space agencies are not interested in making space accessible to taxpayers. They have never tried to do so. For example, the previous NASA administrator tried to persuade Congress to allow NASA to spend $14 billion to develop a capsule like the Apollo module, which they called an "orbital space plane." Even if it was reusable, this would have cost $14 billion, plus at least $100 million per flight, since it would have been launched on a US-made expendable rocket. Thus it would be some ten times more expensive than Soyuz, of which the total cost of rocket and crew-cabin is some $28 million and never as reliable. It would produce the truly bizarre result that more than 50 years after Gagarin's flight, and having spent more than $1 trillion, the US government would have launch costs 10x higher than Gagarin's! Economically the "orbital space plane" is nonsensical; it would prevent the achievement of low-cost space travel in the US for another 20 years. In truth, it was a political project, designed to make expenditure on the International Space Station project sound a responsible use of taxpayers' money in statements made by NASA officials and politicians for consumption by the general public, who are poorly informed victims of this myth.

OECD space agencies' favorite activity is performing government projects, using equipment they have developed. But they are also legally responsible for commercial space development; for example, NASA is required by Federal law to "encourage, to the maximum extent possible, the fullest commercial use of space." But NASA is not doing this—and nor are ESA JAXA. Out of the $20 billion which they spend every year on non-science activities, essentially none is used to forward the development of passenger travel, despite having acknowledged its unique commercial potential. The politicians who allow this situation to continue are not fulfilling their responsibilities to the public.

MYTH 3: "SPACE TOURISM HAS NO ECONOMIC VALUE"

Space agency staff often claim this, but now that applications satellites are a mature business, there is nothing more valuable to be done in space than to make it possible for the general public to travel there. How can Americans believe anything else? Historically this was the source of America's strength: wealth was created by vigorous, consumer-oriented businesses that over-powered the Soviet Union, not military prowess. With the commercial space industry shrinking for lack of demand, space agencies are in denial about this. But the goal of economic development is freedom, the freedom to do what we want. Most people, once they reach a certain standard of living, like to travel, which is one of the greatest educational activities: "travel broadens the mind." Everyone who has been to space says that it was the greatest experience of their life, and market research studies show that a majority of the population in all countries surveyed so far say that they would like to take a trip to space.

In democratic capitalistic countries, no other justification is necessary. It should be sufficient reason that many people wish to pay for this life-enhancing experience. In addition to being unique fun, viewing the Earth from space, looking out beyond toward the beckoning stars, is a profoundly educational and spiritual experience. Not only is this wish to travel to space and to the Moon not "trivial," it is profoundly human and highly desirable for as many people to experience as possible. However, as it happens, R&D in the aerospace industry is nearly all government-funded, and so without some effective popular pressure being put on governments to facilitate the development of this activity, many more years are likely to be wasted, at great cost to taxpayers, as discussed below.

Through 2004, space agencies have spent about $1 trillion of taxpayers' money, with which they have developed a significant amount of space-related technology and knowledge. But much of it is of little economic value, because it is far too expensive. Furthermore, space agencies have made no effort to apply this technology to the most economically valuable use of space—which is to supply the passenger travel services which large numbers of people around the world wish to purchase. Consequently, instead of a $1 trillion/year commercial space industry, there is a commercial satellite services industry with a turnover of around $20 billion/year, which is about 1/50 of what should result from $1 trillion investment. Commercial demand multiplies the economic activity arising from investment by 10x to 20x, as shown in Figure 1. Without some such source of large turnover, investment in space development cannot be repaid, and space commercialization is impossible.

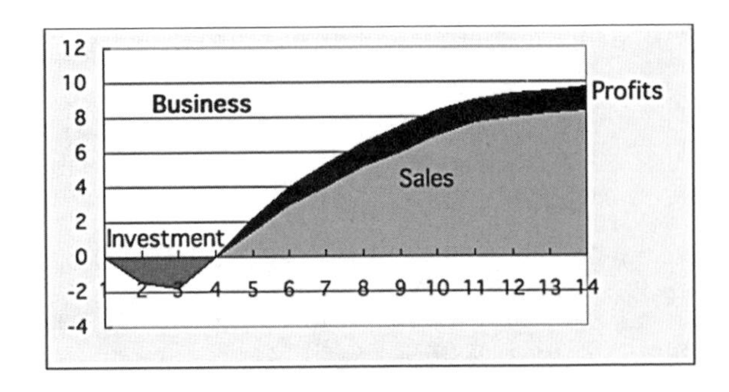

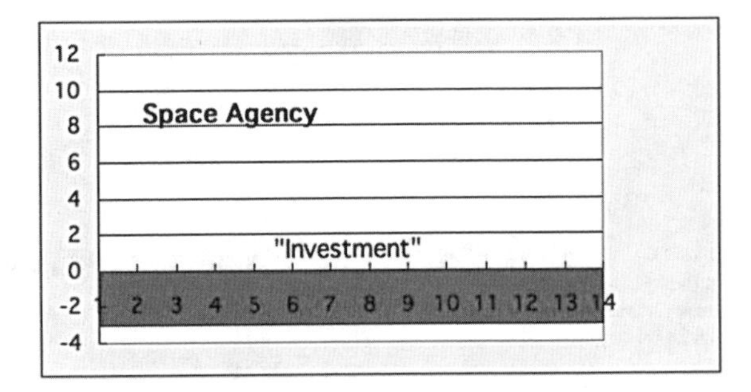

Figure 1: Contrast between commercial investment and space agency expenditure

MYTH 4: "SPACE TOURISM WILL BE AVAILABLE FOR A FEW RICH PEOPLE."

Combined with studies of the potential for cost reduction through airline-like orbital flight operations, this suggests strongly that the business could grow to millions of customers/year. Just as in passenger air travel, as the scale of traffic grows, costs and prices will fall progressively. Based on the work of the Japanese Rocket Society and others, we could have an orbital tourism industry of several million passengers per year by about 2030, as shown in Figure 2, first published in 1999. The great majority of the investment needed would come from the private sector, as in the airline, hotel, cruising and leisure industries today. However, unless some initial investment is provided by governments we will waste many more years waiting to start.

The position taken by heads of space agencies that this is the responsibility of the private sector is disingenuous. Space agencies' economic return on their non-science activities is close to minus 100%. Private investment in a novel activity such as space tourism will require a compound return of some 25% or more. Although the available evidence suggests that passenger space travel will have as great economic value as passenger air travel, it cannot be confidently predicted that investments in the early stages will earn such a return, particularly while there are major regulatory uncertainties due to governments' delay in this matter.

Furthermore, it is economically irrational for governments to spend heavily on space agencies' loss-making activities while refusing to invest in much more economically promising ones. Such a position is not only a "double standard", but it is deeply flawed as economic policy, preventing the growth of a major new industry. Moreover, almost no major aerospace developments are privately funded, as mentioned above, and governments have invested heavily for decades in aviation developments which have had positive indirect benefits that greatly exceed the profits earned directly by airlines (which are quite limited).

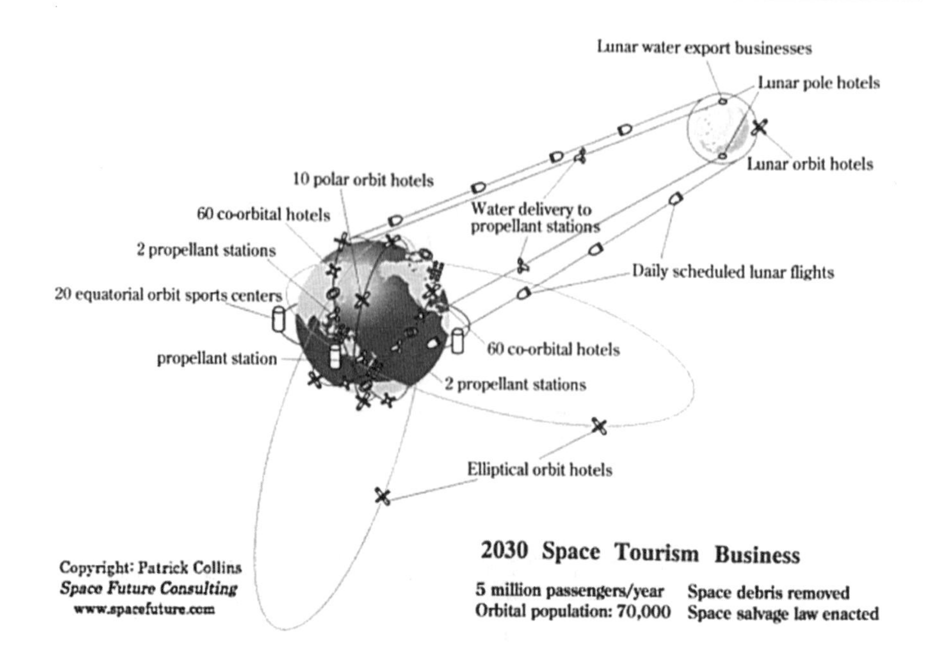

Figure 2: Feasible space tourism scenario

Not only does it seem likely that space tourism can grow into a major new source of high quality employment comparable to air travel, but it requires only very modest government support to initiate. With a budget of just 10% of government support for space agencies devoted to stimulating the growth of space tourism, we could realize the scenario in Figure 2, since most of the later investment will come from the private sector. Funding of $2 billion/year would allow development of both SSTO VTOL and TSTO HTOL passenger vehicles, as well as a range of related activities. Once passenger vehicles are certified for passenger carrying, private companies will take over—just as they manufacture and operate airliners and hotels. In this case lunar tourism could start as early as the 2020s.

In this context it is useful to remember how fast companies can invest when they anticipate profits. Within just a few years, US companies invested nearly $1 trillion into new fiber-optic networks during the late 1990s; unfortunately, they collectively overestimated the demand by more than a factor of ten. However, spread over 30 years the investment of several hundred billion dollars to expand space tourism to millions of passengers/year is clearly easy in comparison.

As the number of hotels in Earth orbit grows following the scenario in Figure 2, it will create a large, near-time market in low Earth orbit for a range of goods and services at high prices. Of particular interest in the early stages will be the demand for water and liquid oxygen (LOX), which could be lunar exports. Even at relatively low launch costs of, say, $200/kg, water in orbit will still be very expensive—$200,000/ton, giving lunar producers a good business opportunity which could grow to billions of dollars/year. Investment in the infrastructure needed for living on the Moon could therefore be repaid largely from exports of ice, water and/or LOX. This infrastructure can then be expanded for tourist accommodation. Once started, the hotel industry, which is a large, powerful and competitive business, will progressively improve and expand their offerings on the lunar surface. Investments in terrestrial hotels costing $1 billion and theme parks costing several billion dollars show that, once launch costs are reduced and infrastructure developed, the scale of financing required to build lunar hotels should be feasible.

In this way the development of tourist accommodation services in Earth orbit can help to trigger the development of lunar tourism services. If passenger flight services to low Earth orbits start within 10 to 15 years, lunar tourism could start within 15 to 20 years, though lunar orbital trips could start earlier. These estimates contrast sharply with space agencies' suggestions that LEO tourism could start in 2040 or even in 2075! It is perhaps worth emphasizing that there is no question of government subsidies for tourism to and from the Moon; this activity will develop on a commercial basis once low Earth orbit becomes readily accessible.

As proved 34 years ago, a return journey to the Moon takes one week. Consequently, ten-day to two-week lunar trips, which are very convenient for tourism, can clearly become popular services. Long-term lunar residents will need to live and work underground for much of the time in order to limit the radiation they receive (although counter-measures can be expected to be developed progressively). However the situation is different for guests, who will be able to spend several "day-times" on the surface without significant risk to their health. As lunar infrastructure develops, the possibility of constructing buildings six times taller than on Earth provides extraordinary potential for fascinating architecture. Domes will be able to be even larger due to the support from the internal air pressure. Consequently, as soon as visitor numbers justify it economically, developers will construct a lunar flying stadium for this new sport. Humans have such a deep interest in flying like birds that the possibility of really doing so in the lunar gravity field will surely lead many people to try it out. Flying will also be possible in orbiting zero-gravity facilities so guests will be able to try it out there first (though it will be significantly different from flying in the Moon's 1/6 gravity field).

The following is a simple list of some of the highlights of a lunar tourist trip, at a time when lunar travel and tourism are relatively mature. In the early days, lunar visits will be primarily a form of "adventure tourism"—but as the infrastructure grows they will offer a progressively wider range of increasingly interesting activities.

1) Preparation— most guests will probably have previously made sub-orbital and orbital flights.
2) Boarding orbital ferry
3) Take-off
4) Entering low Earth orbit
5) Rendezvous and docking at LEO hotel
6) Disembarking, short stay in LEO hotel
7) Boarding Earth-Moon inter-orbital ferry
8) Undocking, departure from Earth orbit
9) Views of ever-shrinking Earth and ever-growing Moon
10) Earth-Moon Libration point 1—58,000 km to go
11) Earth-Moon gravitational equivalence point—38,000 km to go
12) Entering low lunar orbit
13) Rendezvous and docking at lunar orbit hotel
14) Disembarking, viewing Moon from low orbit
15) Boarding lunar surface ferry
16) Departure, de-orbit
17) Landing, disembarking on lunar surface
18) Check-in, acclimatization, "Moon-walking"
19) Sightseeing, views of Moon and Earth
20) Visits to dark side, polar mines, historic sites
21) Flying stadium, flying sports
22) Performances, flying ballet
23) Lunar park, low-gravity pool
24) Check-out, take-off for return journey—reverse sequence to above.

From the above it seems clear that, far from being a trivial activity involving just a few rich people, which would waste resources that could be used better here on Earth, lunar tourism can grow into a large industry which will, on the contrary, have great benefits for the world economy.

Space agencies spend huge budgets on activities that generate no new commercial demand, because they consider consumer demand to be "trivial". But this is the opposite of the truth: without stimulating popular demand, space activities are just a burden on taxpayers. As discussed above, OECD space agencies have spent $1 trillion on nominally civilian space activities to date, but this has created commercial turnover of only about $20 billion/year—about 1/50 of the amount that commercial investment of the same amount would have achieved. From the economic point of view this performance is terrible; it is a pitifully bad use of resources, ranking with the worst of public sector expenditure, and it has added materially to the burden of government debt weighing down OECD economies.

It is very unfortunate that, because of this poor economic performance of space agencies and their political overseers, very few people in government understand the potential of space activities to contribute to economic growth. As the manufacture of clothes, bicycles, electrical goods, ships, televisions, steel, computers and software and other work move to lower-cost countries, now including India and China, there is a "jobless recovery" in the USA. There is a great deal of discussion about whether this flow of jobs abroad is good or bad for the USA. The truth is that it can be good for the US people on one condition: that the jobs lost are replaced with better, higher-paying jobs. There is only one way to do this: continue the pattern that has been operating for centuries whereby the richer countries develop new industries progressively which replace jobs lost in older ones.

If space will learn from aviation and follow the precedent of passenger air travel, there is every reason to foresee that passenger space travel can grow to large scale. The growth of space tourism can produce an ever-widening boom spreading through the economy like the "Internet Boom" of the 1990s—except that this boom need not end in a stock-market crash due to insufficient profits. The demand for space travel has been pent up for more than 30 years, and can grow literally without limit.

Critics sometimes mock advocates of space tourism by saying that "they want everyone to fly to space—which is crazy". But this is a ridiculous claim: the wealth generated by European investment in the Americas did not flow only to the small proportion of the population who crossed the Atlantic—much of it benefited investors who stayed behind in Europe. The grand old houses on the west coast of Britain in Bristol, Cardiff, Liverpool and Glasgow are testimony to the wealth of merchants who traded successfully with America without leaving Britain.

The same will happen on Earth as lunar tourism develops—just as the growth of Hawaiian tourism has generated ever-growing trade with the USA and Japan, stimulating investment in new aircraft, hotels, computer systems, and many new sports and leisure industries, among other products and services. Just a few million people per year visiting orbit will have a highly stimulating effect on a wide range of aerospace-related industries which are today shrinking from lack of demand. Yet air travel is already three million people per day, so space tourism has plenty of room to grow. By becoming a large new industry which will purchase a wide range of advanced technologies from a wide range of engineering industries, space tourism will create profitable, expanding employment for millions of people on Earth, quite different from those who actually choose to visit or work in space and on the Moon. Lunar tourism will raise humans' eyes to a wider cultural horizon. People still speak of the "post-cold-war" era or the "era of globalization"—but these are merely transitional phases. Tourism in space is going to lead humans into the true "Space Age"—that is, the age when the general public starts to make use of space directly. This will bring about a genuine "New Renaissance"—a time when old ideas that have been accepted for generations are understood to be wrong, and a flood of new understanding revitalizes stagnating cultures in an optimistic era of intellectual excitement. Ideas that are long overdue for discarding are that space flight has little value and can only be justified for a small number of government employees, and that humans face a constrained and gloomy future of fighting over "dwindling resources."

In a link with the original Renaissance, this will herald the start of the true "Copernican" era. The idea is still widespread that here on Earth is all that humans have; energy sources are buried in the ground, and so we need to grab as much as we can before others do. But this idea is so narrow-minded it is laughable; people who believe this are the "Flat-Earthers" of the 21st Century. If these people will just look "outside," they will learn that the resources of Earth are barely a speck compared to what is available to humans in reality.

But while "Flat-Earthers" continue to hold sway in governments around the world, and space agencies expend enormous amounts of taxpayers' funds with little economic benefit, just a modest investment would be sufficient to develop inexhaustible energy resources within a few decades. This will permanently eliminate any need to burn limited fossil fuel resources, and will include the development of space-based power supply to Earth.

In view of these wide-ranging benefits, there is surely nothing else currently foreseeable for human civilization—not robotics, not broad-band Internet connectivity, not genetic engineering—which offers anywhere near such an inspiring and uplifting prospect for human culture worldwide—a true New Renaissance to which every culture can contribute, and from which every country can benefit—as the rapid growth of tourism in space and on the Moon.

With the growing recognition of the economic importance of space tourism, which advanced rapidly in 2004— we are approaching a watershed in progress towards its realization. However, if voters are so uncritical that they continue to accept government space agencies' self-interested statements about the infeasibility of space tourism until 2040 or even 2075, then they will continue to bear the heavy cost of closing off the enormous promise of space commercialization.

As soon as the effort is made to compel governments to start spending the small amount needed to facilitate the development of space tourism, just 10% of existing space budgets earmarked for this purpose will change the world. And if the politicians responsible for space agency budgets will not compel them to fulfill their already clearly defined responsibilities in this field, then it will be necessary to fund new government organizations to break the space agencies' extremely damaging monopoly position. To do this the public must be educated about these possibilities—as the X-Prize helped to do. Once the public learns about the possibilities that space agencies are preventing, we know from market research that they will support the changes necessary to realize them. There is a major role for private patrons in this, for example through sponsoring university-based research.

Once we start on this path, our progress will be a beacon for the young, showing them that—in contrast to the ridiculous "Flat-Earthers'" vision of rich countries having to fight to the death to keep the poor from getting their share of the "dwindling resources" of Earth—there is an easily-achievable, optimistic future of unlimited growth in standards of living for everyone on Earth, in orbit, on the Moon—and beyond.

And just as soon as it's accepted that developing orbital tourism is of vital economic importance, the road to lunar development will be open—and it will not close a second time.

Patrick Collins is currently Professor of Economics at Azabu University in Japan, and Director of Space Future Consulting, he was born in Sussex, England in 1952. He graduated from Cambridge University; earned an MBA from London University, and a PhD on the economics of space-based solar power supply in 1985. The major focus of his research at universities and space research centers over the past 25 years has been the potential for commercialization of space activities. Having identified tourism as the most promising market in space, Professor Collins has become the leading economist in the field, having produced over 100 publications on the subject. Professor Collins considers the missed opportunity, whereby the "cold war" space industry has spent $1 trillion without reducing the cost of flying to space by a single cent below the "Soyuz" rocket, one of the most costly—and continuing—mistakes of economic policy over the past half-century.

Dennis Wingo is one of those guys that make any establishment nervous. He comes across as an easy going, laid back southern boy, complete with a drawl and faded jeans, yet his mind is the match of anyone to ever cross the threshold of an Ivy league campus. Worse, his ideas challenge the status quo, and worse yet, his grasp of the science and data to back his assertions make him hard to ignore. Among the rationales for our return to the Moon, the energy question is often tossed on the table, usually in terms of space solar power. But given recent initiatives to begin the development of a hydrogen based economy, the need for platinum, which is central to the workings of hydrogen engines, has provided yet another rationale, one which might trump all the rest in its potential value. Aside from the direct point Dennis makes in the following essay, there is a more subtle, yet powerful thing to consider. Two decades ago no one rally had considered the Moon might hold the key to an energy revolution on Earth. Two decades from now what other new and important reasons might we find for being there? What other resources, wealth or possibilities might present themselves, especially if we are there, living, working, experimenting, testing, surveying, mining, and developing what we find just outside the airlock of our homes, labs and workplaces, as opposed to having to speculate and guess based on the sporadic and almost random data of this or that short lived "scientific" mission? What if we went to stay, and we did it on purpose? Think about it. Like the rest of our authors, Dennis has.

Asteroidal Resources and the Cis-lunar Industrial Economy

by Dennis Wingo

Background

Today we live in interesting times with our civilization reaching the peak of oil production, instability in areas where most of the planet's oil reserves are, and the rise of economic powerhouses in China and India. Unless space exploration, and the return to the Moon, can become part of the solution of the problems that confront our civilization its value is limited. Fortunately it is the case that the resources of the Moon, asteroids, and our solar system are so much greater than on the Earth that it is imperative that our goals for space exploration have resource development as the central theme.

In my book "Moonrush" that theme was the near term development of resources on the Moon, namely the resources that are derived from the impacts of metallic asteroids on the Moon. While the emphasis by many has been on the potential for water in the permanently unlit craters at the lunar poles, that water will be best used to support the exploration and development of the metallic resources of the Moon. It is well known in the scientific literature that approximately 3% of inner solar system bodies are made of an alloy of Nickel, Iron, Cobalt, and an assemblage of Platinum Group Metals (PGM's).[1]

The reason for an emphasis on PGM resources is the fact that Platinum is the crucial element for the construction of fuel cells. Fuel cells are the crucial underpinning of the hydrogen economy and the hydrogen economy is the basis for a transition from a petroleum hydrocarbon economy to a sustainable, non CO_2 generating energy source, for maintaining and even increasing the level of affluence of our global civilization.

The Hypothesis

The operative hypothesis that has been developed is that by knowing the statistical distribution of asteroidal types, (the 3% M class "irons") we can then look at the Moon for near term resource opportunities with this type of asteroid. It is clear that large iron asteroids have impacted the Earth in the past. Figure 1 shows the famous Sudbury district in Canada and the location of copper, nickel, and PGM mines:

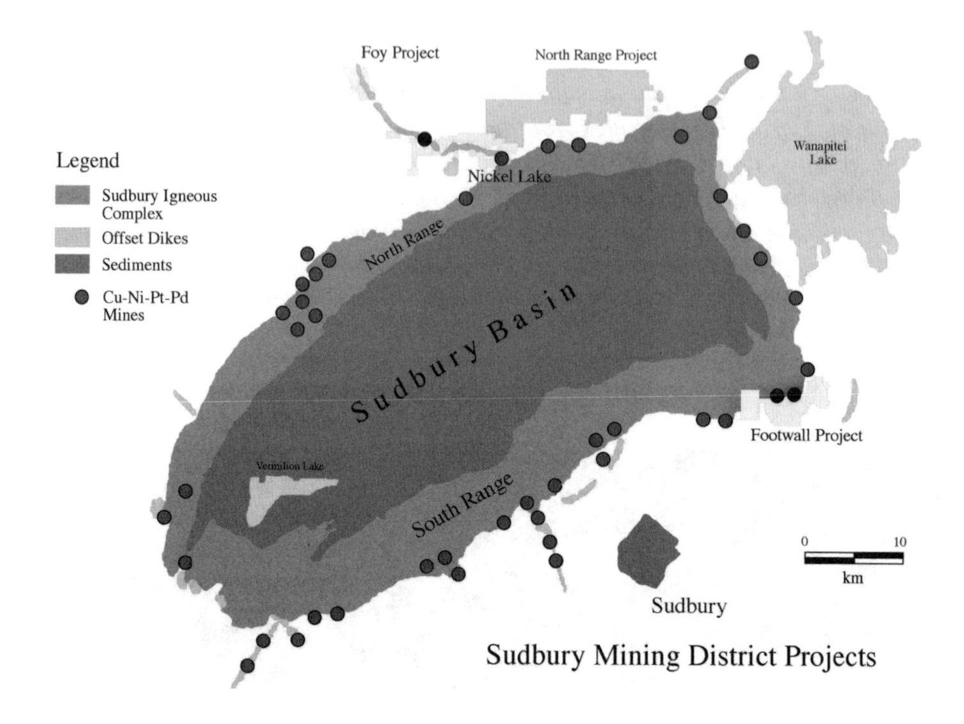

Figure 1: The Sudbury Mining District

The Sudbury district is considered by many geophysicists to be an "astrobleme", which means star wound in French. This theory was first developed by Dr. Paul Dietz in 1963 and the in essence states that the Sudbury basin is the remnant of an M class asteroid impact approximately 1.9 billion years ago.

The Sudbury area is the location of one of the world's great nickel, copper, and PGM mines and the Canadian government estimates that the total resources from this site has a value in excess of 300 billion dollars.

Another great impact that is relevant on the earth is the Vredefort impact site in southern A two billion year old impact site where the Bushveld complex containing the Merensky reef is located, the source area of over 70% of the known planetary reserves. Figure 2 shows the impact site:

Figure 2: Veredfort Impact Astrobleme (Picture Courtesy NASA)

One very good paper written by Dr. Wofgang Elston discusses the possibility that the vulcanism that produced the Bushveld complex was a result of the Vredefort impact.[2] There are also hints that two sites in Russia where most of their PGM's are mined are from nearby impacts.

Meteorite Data

In a 2004 paper by Petaev and Jacobsen the fractional composition of several M class iron meteorites were defined with great precision.[3] PGM concentrations in iron meteorites are comparable to the quality of ore at the Merensky reef (9 to 18 grams/ton) and at Sudbury (3-9 grams per ton). Table 1 shows a sampling from the paper of the concentrations of Platinum (Pt), Ruthenium (Ru), Iridium (Ir) and Gold (Au) in ppm.

Meteorite	Pt	Ru	Ir	Au
Arispe	22.85	3.008	11.06	0.771
Bennett County	38.89	26.93	43.87	0.419
Cape York	11.85	7.47	3.13	0.848
Henbury	18.31	14.04	13.78	0.422
Grant	2.84	1.85	0.01	1.910

Table 1: Sample of PGM Concentrations in Selected Meteorites

There is considerably more data in the paper but the above samples give a flavor of the distribution of PGM concentrations in the samples. Their paper was on the analytic techniques for PGM fractions in meteorites so the data has a high level of precision. It is interesting to note that in many of the samples the PGM concentrations are greater than in comparable terrestrial resources. How does this tie to the Moon? With 3% of all impactors being M class bodies, if it can be shown that any significant quantities of these bodies survive impact then we have the potential for a very valuable resource for both lunar and terrestrial applications.

Impacts on the Moon

Anyone that looks up in the sky on any night that the Moon is up realizes that it has been whacked by a lot of asteroids. As the Moon is part of the inner solar system, and since most asteroids are from the main asteroid belt it is not much of a stretch to say that the distribution of impacts on the Moon and their makeup should mirror what we know about the asteroid belt and its composition. In order to determine the survivability of any one object on the lunar surface you have to know three things, one the impact velocity, two the material strength of the impacting object, and three the strength of materials of the place where the body impacted. Figure 3 gives the impact velocity distribution for objects hitting the Moon:

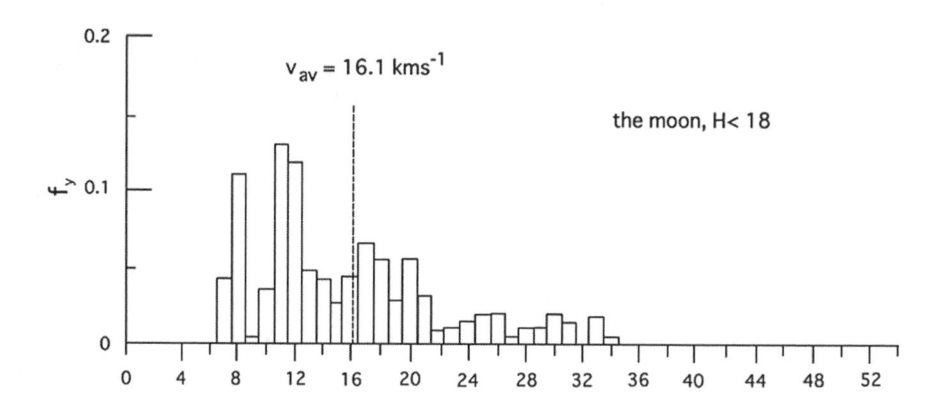

Figure 3: Lunar Impact Velocity Distribution

The chart above is from a paper by Neukum, Ivanov, and Wagner.[4] This paper is the upper bound on the average impact velocity. Other work by NASA[5] indicates that the average impact velocity is closer to 14 km/sec. The two papers give the bounds for the average of all impactors on the Moon with the low end being at about 6.2 km/sec. Note that in the chart in figure 3 the largest numerical amount of objects are somewhat below the average with a large distribution of faster objects making up the high side.

Now let us look at the distribution of the size of impactors on the lunar surface versus frequency. In the four billion year lifetime of the Moon, millions of asteroids, large and small, have impacted the Moon. There has been little tectonic activity to resurface the Moon, as has been the case here on the Earth. What little resurfacing there has been is starkly visible in the form of the Mare regions. The number of craters on the Moon is directly related to the age of the surface. The highland regions of the Moon are 3.8 to 4.2 billion years old and therefore have the greatest density of craters. The lowlands are 3.1 to 3.8 billion years old and have fewer craters. There should not be that much difference in the number of craters, but there is, leading to the postulation of a "heavy bombardment" period during the period just after the formation of the lunar highlands. Figure 4 illustrates the frequency and size distribution of highland and Mare craters:

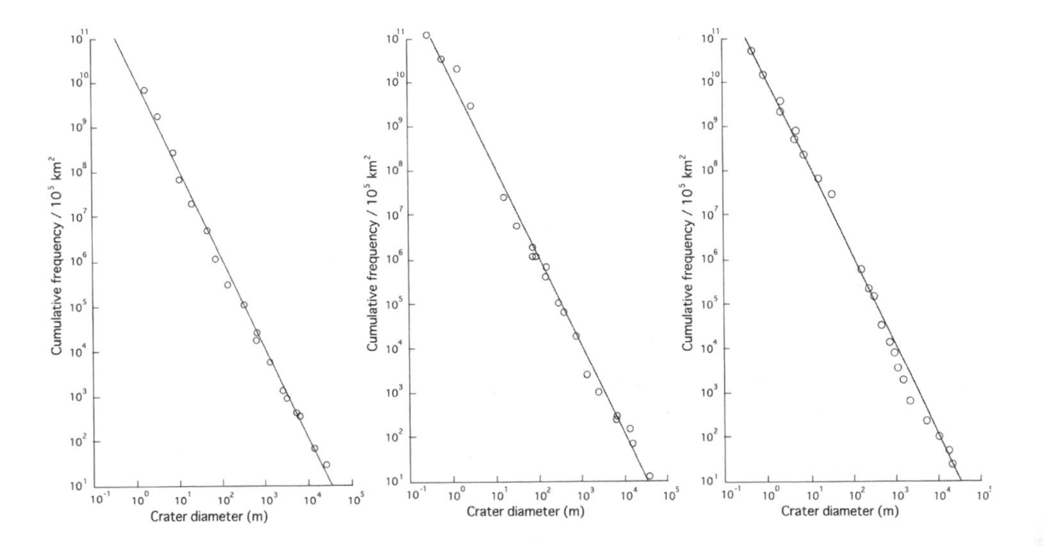

Figure 4: Impact Crater Frequency, Mare Tranquillitatis, Nubium, and Alphonsus

In the three charts of crater frequency vs. size, Mare Tranquillitatis has the fewest number of craters, implying the youngest age. Mare Nubium has 1.4 times more craters per unit area, and the Highlands region of Alphonsus has 2.5 times as many impacts per unit area than in Tranquillitatis, making this the oldest region of the three. The number of craters 1 km in diameter or larger for the three regions are:

Mare Tranquillitatis	10,000 Craters per million square kilometers
Mare Nubium	14,000 Craters per million square kilometers
Alphonsus Region	25,000 Craters per million square kilometers

This data was derived from three of the early lunar impacting spacecraft, Ranger 7, 8, and 9. Dr. C. A. Cross examined the pictures from these three spacecraft and developed an inverse power law with a slope of –2 that allows for a mathematical extrapolation to allow derivation of crater frequencies and sizes outside of the resolution of the Ranger images.

From this information some gross generalizations can be made. Since the total area of the Mare on the Moon is approximately 19% of the total surface area, and the total surface area of the Moon is approximately 38 million square kilometers, the number of craters larger than 1 kilometer in diameter is about 86,400 in the Mare regions and 845,000 in the highlands regions of the Moon. Of these impacts, 3% are M class metal asteroid impact scars. This means M class impactors make up about 2,592 impacts in the Mare regions and 25,350 in the Highland regions of the Moon. For comparison, a 1 kilometer impact is an object about the size of the Canyon Diablo impactor (Meteor Crater Arizona), which was a metal asteroid about 15 meters in diameter. It weighed nearly 100,000 metric tons and would have contained 1-10,000 kilos of PGMs.

With the total number of M class metal impactors in the range of ~28,000 objects of the same general size as the Canyon Diablo impactor, this works out to be a lot of metal. If at the absolute minimum, all of the impacts of M class objects were the same size as the Canyon Diablo impactor, the total amount of metal would be 3 billion metric tons, having 62 million kilos of PGMs (assuming 20 grams per ton average PGM concentration), 1.2 times the total amount considered commercially viable to extract on the earth by the South Africans. In truth, the amount is probably a 1,000 to 100,000 times that amount based upon the scaling law derived by Cross as shown in figure 4. However compelling this first brush thought experiment seems to be, the reality is a little different.

Now that we have the average velocity we can look at how the material strength of an Ni/Fe asteroid, along with the impact velocity as described, act together as two parts of the survivability equation. In a definitive paper by Schnabel, Pierazzo, Xue, Herzog, Masarik, Cresswell, Tada, Liu, and Fifeld, in volume 285 of Science in 1999, the authors looked at the dynamics of the impact of the Canyon Diablo nickel/iron asteroid. They took a unique approach that mixed hydrocode modeling and experimental determination of the rare isotope nickel 59. Hydrocodes are numerical integration programs that compute the propagation of shock waves, velocities, strains and stresses, as a function of time and position. The nickel 59 isotopes have a half-life of 76,000 years and are created when the asteroid is in space, subject to radiation from high-energy cosmic rays. There is a relationship between the production of nickel 59 and nickel's depth inside the asteroid, governed by the shielding effect of the outer portion of the body.

By using the appropriate measurement equipment, samples of melted spherules and metal meteorite fragments from the Canyon Diablo impactor were measured. The results of the measurements showed that the solid fragments from the impactor had seven times the nickel 59 of the spherules that had melted. What this means is that the solid fragments were from the outer edges of the impactor and the once liquid spherules originated deeper in the asteroid. In their hydrocode modeling, the scientists modeled the impactor as a 15-meter spheroid impacting at either 15 km/s or 20 km/s.

In comparing their modeling to the actual amount of body estimated to have been recovered by an early researcher on the Canyon Diablo impactor (Ninninger) (4000-7500 tons of spherules and an undetermined amount of solid fragments due to removal before Barringer's ownership of the crater), they determined that the impact velocity was more likely 20 km/s. They estimated that a shell 1.5-2 meters thick, constituting ~15% of the asteroid, remained intact! Figure 5 shows the graphic of their hydrocode simulation of the Canyon Diablo impactor.

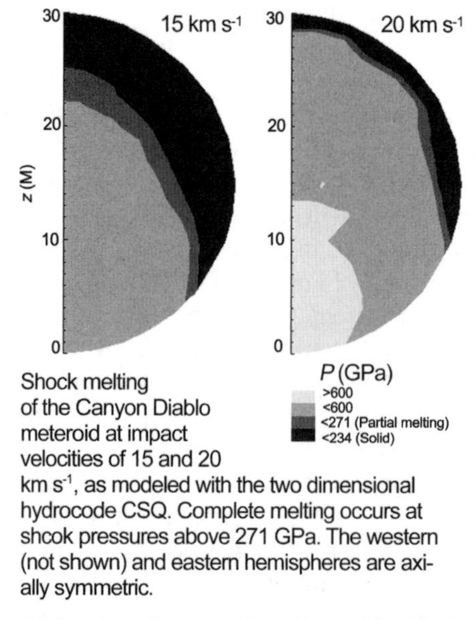

Shock melting of the Canyon Diablo meteroid at impact velocities of 15 and 20 km s^{-1}, as modeled with the two dimensional hydrocode CSQ. Complete melting occurs at shcok pressures above 271 GPa. The western (not shown) and eastern hemispheres are axially symmetric.

Figure 5: 15 km/sec vs 20 km/sec Survivability Based On Hydrocode Modeling

For a lower velocity of 15 km/s, they estimated that the solid shell would have constituted 63% of the body that would have remained solid. Since this is not consistent with the evidence, their estimate of the impact velocity is closer to 20 km/s.

Interestingly and importantly for the purposes of this book, at neither velocity did any part of the asteroid vaporize because the peak shock pressures did not reach the threshold for incipient vaporization of 800 giga-Pascals. The empirical evidence from the spherules is that they came from a depth of 1.3-1.6 meters beneath the surface of the impactor. This is consistent with the hydrocode modeling. These are among the definitive results of the paper that are applicable to the hypothesis concerning M class impactors on the Moon.

Interestingly and importantly for the purposes of this book, at neither velocity did any part of the asteroid vaporize because the peak shock pressures did not reach the threshold for incipient vaporization of 800 giga-Pascals. The empirical evidence from the spherules is that they came from a depth of 1.3-1.6 meters beneath the surface of the impactor. This is consistent with the hydrocode modeling. These are among the definitive results of the paper that are applicable to the hypothesis concerning M class impactors on the Moon.

Let us now extrapolate this to lunar impactors of the M class metallic type. Referring back to figure 8.5, the average impact velocity of impactors on the Moon is between 14-16 km/s. This is at the lower end of the impact scale for Canyon Diablo. Therefore, using a very conservative approach, it is reasonable to expect that between 15-63% of the mass of the estimated 28,000 big impactors (those that make a crater at least 1 km in diameter) would be preserved in the general vicinity of the impact. Furthermore, there would be zero oxidation of any of these impacts due to the lack of oxygen.

Fortunately, this is not the whole story. Looking back at figure 3, you will see that the distribution of impact velocities extends down to a minimum of about 6.5 km/s. It is reasonable to assume that, with lower impact velocities, the survivability of these M class metal asteroids is higher. One bit of evidence to support this comes from a paper titled, *"Meteorite Accumulations on Mars."* In this paper, the researchers used a hypervelocity gun to shoot objects of various types into material that simulates the Regolith of mars and the moon. With their system limitation of a maximum velocity of ~2 km/s and their focus on Mars, they still obtained data of interest for our lunar focus. They did fire small nickel/iron meteorite fragments into the simulated Regolith at 2 km/sec and the result was deformation of the object but no further fragmentation.

There is another more recent discovery that has a bearing on the subject of Ni/Fe impactors. On January 20, 2005 the NASA Mars Rover Opportunity, scouting the surface of Mars for interesting rocks found the first confirmed discovery of a meteorite on another planet. This meteorite, weighing several hundred kilograms and the size of a basketball was found simply sitting on the surface of the red planet. Figure 6 shows this meteorite:

Figure 6: Mars Nickel/Iron Meteorite

Therefore, we have to separate confirmed sets of data from different researchers as well as visual evidence from the surface of Mars. We have the results of the hydrocode modeling of Pierazzo and company, coupled with the hypervelocity data from Bland and Smith. This presents a gap in the data between 2 km/s and 15 km/s that is ripe for investigation. Betty Pierazzo has indicated in personal communication that they do have data down to 10 km/s, but this gap requires more intensive study to determine the survivability of M class impactors. One can reasonably postulate that the survivability increases as velocity decreases and that the possibility exists that, at the low end of the known impact scale as shown in figure 4, there may be some pretty, darn big chunks of nickel/iron/cobalt/PGM objects laying on the lunar surface.

Using the absolutely conservative assumption that none of the asteroids are larger than the Canyon Diablo impactor's weight of 100,000 tons and assuming the average impact velocity of 15 km/s, we have between 450 million to 1.77 billion tons of economically recoverable nickel/iron/cobalt/PGM material in the vicinity of the impact site. Applying another conservative estimate that more realistically represents the known size-frequency curve as shown for the highland region in figure 8.3, we get nearly 1300 craters that are 10 km in size, resulting in 39 impacts of asteroids that are more like 150 meters in diameter. Taking this to the next level, we have 100 craters that are 100 km in diameter, resulting in 3 M class metal impactors that would be 1.5 km in diameter, about the size of the $20 trillion dollar 3554 Amun asteroid. This takes the resource base up to between 140-590 billion tons of localized, recoverable nickel/iron/cobalt/PGM impact material. This is on the same scale as the estimate by scientists concerning water trapped in the lunar polar region.

All of the above calculations ignore that another 1% (of the total of all impacts) of all impactors are 50-50 stony-nickel/iron bodies or that another group of asteroid impactors are Carbonaceous Chondrites that have highly elevated PGM concentrations. Even in stony asteroids, there are a significant percentage of nickel/iron inclusions in many of them. The total amount of nickel/iron/cobalt/PGM materials on the moon is easily several times that of what is discussed in the conservative model I show here.

The only issue is that these bodies are more easily disrupted, pulverized, and splattered across the face of the moon since they are not as strongly held together as the M class metal bodies. Another factor that positively influences maximum pressures and the resulting behavior of impactors is the impact angle. The previous modeling studies referenced here assume a vertical impact angle, which is unlikely. There are also differences in the fate of the impactor depending on what type of material that they slam into. Those who slam into deep regolith versus igneous rock behave differently. The character of the lunar material also affects the explosive rebound and distribution of the rest of the body that hits the Moon. The results of this more global distribution of impact products are well known due to the Apollo program.

Global Distribution of M Class Impact Material

The Moon has a total surface area of ~38 million square kilometers. If you divide this by the total number of M class metal asteroid impacts (~28,000), you get a statistical average of approximately one 15-meter size impactor per 1357 square kilometers. For larger objects of ~150 meters (approximately 39 of them), this works out to one per 974 thousand square kilometers. For the largest objects of ~1.5 km, this works out to one per 12 million square kilometers. If you look at the velocity distribution curve in figure 8.5, the bias is well toward the lower velocities. This is especially true of asteroids because they all are in the lower velocity group, whose orbits evolved from the main belt and are now not that different from the orbit of the earth.

Testing the Hypothesis

All of what is written in this chapter is a hypothesis. A hypothesis must be testable in order to begin the move from a thought experiment to a resource that can be used by lunar explorers and our industrial society back on Earth. It should be noted that our knowledge of the lunar surface is not very good. The best resolution in the visual band on the Moon is in the equatorial regions photographed by the first three of the five lunar orbiters. Further imaging was accomplished by the Apollo astronauts who were left in the Command Service Module (CSM) during the lunar landings. These images were able to see many boulders, but with no spectrographic data at these resolutions nothing supporting the Ni/Fe impact hypothesis is gleaned. The Clementine, Lunar Prospector, and now the SMART-1 spacecraft have multispectral imagers but their resolution is in the tens to hundreds of meters. These imagers are not really calibrated for the particular task of finding these impact products but further research might enable such detections. Following is a set of new instruments that would definitively find such bodies should they exist on the lunar surface.

High Power Radar

One very good method of discovering these bodies would be a multi-hundred watt to kilowatt class short wavelength (1-20 cm) radar imager. The Ni/Fe bodies should have a much higher reflectivity than adjacent material and would show up as bright patches in any processed data. With high enough power and a short wavelength we should be able to build maps of the distribution of fines, such as were found in regolith samples brought back by the Apollo crew. It should be noted that the Apollo landing missions did find Ni/Fe products, it makes up between 0.1-to1% by weight of almost every sample of regolith.

Infrared

Another very good method of detection would be to fly an infrared camera (1000-5000 nanometers wavelength) that would have a few meters resolution. The spacecraft carrying this payload would fly in an orbit that takes it over the lunar terminator just after sunset. The Ni/Fe bodies, with much greater thermal conductivity, would release any heat stored during the day much faster than the surrounding bulk regolith. This would find large blocks down to fairly small samples.

Magnetometers

During the Apollo missions there were two microsats deployed from the Apollo 15 and 16 CSM that carried sensitive magnetometers. Several anomalous localized magnetic fields were discovered that are at least somewhat inconsistent with most theories regarding lunar formation. There is at least a sporting chance that a more sensitive instrument, flying at low altitudes could find some of the larger bodies, that due to having penetrated to a depth well below the existing regolith boundary, would be found in this manner. Imagine if you will for a second at least the possibility of an impacted body kilometers in diameter! This would completely change our entire focus for exploration as the resources involved would be orders of magnitude greater than anything on the Earth.

Gravity Measuring Satellites

Multiple spacecraft in known orbits with super stable RF beacons, along with receivers on the Earth and the spacecraft could be used to greatly improve the gravity map obtained by the Lunar Orbiter, Apollo, and Lunar Prospector missions. This would be the only way to improve the gravity map of the Moon's "lumpy" gravity field on the lunar farside. With enough accuracy, localized gravitational anomalies, some already found on previous missions, could be better charted and would be then correlated to data obtained by previous missions.

The above set of instruments together should definitively answer the question and confirm or refute the hypothesis described in this chapter. They would also in a more generic sense greatly improve the state of lunar science in general and help to answer many of the questions related to the formation of the Moon. Imagine the "Moonrush" should this hypothesis become confirmed and its impact on our future.

The "Impact" of Lunar Derived Asteroidal Metals

We already know that there are several inner solar system asteroids with vast quantities of iron, nickel, cobalt, and PGMs. This is not debated. One NEA that we know about is called 3554 Amun. This is one of the ***smallest*** known M class asteroid. It is approximately 2 kilometers in diameter. It also contains more iron, nickel, cobalt and PGMs than have been mined in the history of mankind. Dr. Lewis estimates that the iron and nickel in 3554 Amun is worth about $8 ***trillion*** dollars. The cobalt is worth another $6 ***trillion*** dollars. The PGMs and gold contained in this object adds another $6 ***trillion*** dollars to the total. This is a grand total of $20 ***trillion*** dollars worth of metal, $6 ***trillion*** dollars of that being enough PGMs to completely enable the switch to the Hydrogen Economy with enough left over to adorn every person on earth with a nice PGM ring! While it is unrealistic to expect that a flood of metals on the market would keep the market value this high, the purpose of illustrating these numbers is that the amount of metals obtainable from Amun exceeds the total available Earthly supply. Amun is the smallest identified nickel/iron NEA. The largest that we know of is the main asteroid belt object called 216 Cleopatra. It is 217 X 94 ***kilometers*** in diameter. This massive hunk of iron is larger than Scotland and Wales put together in area and has more PGMs than we could ever possibly use.

Lets say, just for the sake of argument, that we find several small impact fragments that are the equivalent of a 200 meter diameter asteroid. This would bring well over a hundred million tons of iron, nickel, cobalt, and associated PGMs. What would we do with all this stuff? First, it is likely to be found at least within a few hundred linear kilometers of the lunar poles where the water is thought to reside. With water for propellant and metals for industry, the beginnings of a lunar industrial economy are not far behind. Energy would be in the greatest demand, just as it is here on the earth. What if we could land several hundred kilowatts to a megawatt of solar arrays on the lunar surface at the poles and then use that to turn water into hydrogen and oxygen that could run high power fuel cells. (I want to ignore nukes for now)

Having this power available on site would first allow us to melt the body and use it first without consideration of the PGMs to build a railroad to the polar area. With this railroad in place large quantities of the body could be taken back to the poles where the power available there would allow for either the carbonyl or a hot process to be used to separate the nickel, cobalt, and PGMs from the iron. Some of the nickel could be retained to strengthen the iron for use to build large buildings on the lunar surface that could be used for many things, such as agriculture, high technology manufacturing, or habitations, with less than three feet of iron giving the same protection from radiation as the Earth's entire atmosphere. Having large habital volumes is crucially important to start building a human centric infrastructure on the lunar surface.

Between the large habitats could be railroads, monorails or even flat plates on the surface to eliminate the buildup of regolithic dust that is so damaging to equipment on the Moon. A spaceport could be built that would be several hundred meters above the lunar surface where the effluent from their propulsion systems could be captured during a landing cycle. Large industrial machinery could also be built that could be used to landscape the land nearby on a large scale to make it easier to operate equipment and support the emplacement of large, multi-megawatt solar arrays. With an infrastructure in place for power and transportation the large scale processing of the PGM bearing body could commence.

Today platinum on the Earth goes for over 800 dollars per ounce with no reduction of demand in sight. It is estimated that for the hydrogen economy to be viable, production rates for PGMs on the Earth must increase a minimum of five fold, while the price must be reduced by a factor of ten to make fuel cells a viable alternative to the internal combustion engine. This is simply not going to happen on the Earth and the raw production cost today is over $200 per ounce with no dramatic reductions on the horizon. If the lunar installation can process a Ni/Fe body with at least 30 grams per ton of Platinum and other PGMs at a rate of a million tons a year we could produce enough to provide catalysts for 5 million cars per year, assuming that the projected future loading of .2 ounces of platinum per vehicle is achieved. The dollar value for this is not that high at today's prices, being about $850 million dollars but with this as a start, and with the ability to sell the cobalt, which is thousands of times more plentiful but much less expensive is a possibility as well. With a million tons of Ni/Fe produced per year there are bound to be more uses for these metals such as to build large GEO communications platforms and other things that we have not dreamed of yet.

PGMs are not the be all and end all solution to our resource problems but for every kilo that is produced on the Moon that is that much less pollution on the Earth as PGM mining here is a toxic chemical process that also takes huge amounts of electricity and natural gas. We did a study on this and it seems that the importation of 30,000 kg per year of PGMs from the Moon would be the equivalent of beaming a gigawatt of power from the Moon to the Earth, an idea that has been pushed for decades. In reality PGMs derived from an industrial process on the Moon would lower the demand for energy on the Earth. This would enable us to go out and get the more plentiful asteroidal resources and with what we learned on the Moon, increase our production to whatever level is necessary to sustain our civilization and enable the changeover to the hydrogen economy. This is the ultimate goal, to enable us to get beyond the oil demon that threatens to choke the life from the world as it becomes more and more scarce. Matthew Simmons, the founder of one of the world's largest oil investment banks in his book "Twilight in the Desert" recently wrote that trillions of dollars will have to be invested in an energy infrastructure to replace that of oil. It is the position that I have here that this money would be much better spent in developing the resources of the solar system to transcend the limits to growth that relying on an Earthbound solution will ultimately drive us. The choice is ours.

[1] Lewis, J. Hutson, M; *Asteroidal Resources Opportunities Suggested by Meteoric Data*; Resources of Near Earth Space, page 525, University of Arizona Press, Tucson, AZ 1993

[2] Elston, W.E; The Proterozoic Bushveld Complex, South Africa; Plume, Astrobleme, or Both?; Department of Planetary Sciences, University of New Mexico

[3] Petaev, M, Jacobson, S; Differentiation of Metal-rich Meteoritic Parent Bodies; I. Measurements of PGEs, Re, Mo, W, and Au in Meteoric Fe-Ni Metal; Meteorics & Planetary Science 39, Nr 10, 1685-1697 (2004)

[4] Neukum G, et al, *Crater Production and Cratering Chronology for Mercury*, Mercury: Space Environment, Surface, and Interior, 2001 (8027.pdf)

[5] http://cmex-www.arc.nasa.gov/CMEX/data/SiteCat/sitecat2/crater.htm

[5] www.dnp.fmph.uniba.sk/etext/45/text/SciCanDiab.pdf

Dennis Wingo is a 22-year veteran of the computer, academic, and space communities. He worked for early computer pioneers in the development of local area networks which eventually led to innovations such as DSL. He was also an integral force in the use of commercial systems for use in space and flew the first MacIntosh on the Space Shuttle as experiment controller. Dennis received his degree in Engineering Physics at the University of Alabama in Huntsville where he won honors for his academic publications and for his unique approach to small satellite development. Dennis is the Founder & President of SkyCorp Incorporated and has developed a patented approach to the development of highly capable spacecraft manufactured on orbit on the Space Shuttle or International Space Station. SkyCorp has also qualified payloads for flight to the space station via the Russian Soyuz vehicle. Dennis is the author of the book *Moonrush*.

Yoji Kondo is the only author other than myself to appear twice in this collection, the reason is simple, he addresses two important subjects and does so well—and directly. Earlier in this book you read Dr. Kondo's take on the old debate between those who thought we should send people into space and those who want to send robots. Dr. Kondo also has not been shy when it comes to issues more directly related to his field. For example, he was one of the voices that wanted to see the Hubble Space Telescope mounted on a nice solid rock—like say perhaps the Moon—when it was first proposed. As NASA looks to spend billions more on space-based telescopes (and for some purposes space is a great place to put them) Yoji makes the point that the surface of the Moon is a nice solid place to look at the rest of the universe (and indeed back at the Earth). Along with the points he makes, it is important to understand that the building, operation, and maintenance of lunar telescopes will be another of many cumulative activities which we need to make the case for humans to live there. And a valid case it is, so read on...

Astronomy from the Moon

by Yoji Kondo

Introduction

[I]

The surface of the Moon is in vacuum, the Moon provides a solid and inertial platform for telescopes, and its far side—the side that always faces away from Earth—provides radio-noisefree site for radio telescopes. Those are two major and unique assets of the Moon for astronomical work. That is: (a) the Moon is an immense optical bench in vacuum, which is suitable for astronomical interferometry; (b) the far side of the Moon that will shield radio telescopes there from man-made radio noises from Earth. In addition, the Moon is a massive platform from which we can point all sorts of telescopes without requiring (a minimum of) three gyros and momentum wheels for pointing telescopes in free orbit. The types of astronomical observatories that may benefit from the lunar environments are numerous. But what we will build there will depend on what we are planning to do on the Moon. (A) Are we going to the Moon for the purpose of establishing observatories? If that is the case, for practical reasons, we will probably limit our astronomical activities to those fields where the Moon provides the unique opportunities (not available elsewhere in the vicinity of Earth) mentioned at the beginning. (B) Are we going to establish permanent human bases on the Moon for various purposes (not limited to astronomical work), such as a midway station to the rest of the solar system (possibly using solar powered magnetic catapult or for launching nuclear powered rocket that may not be suitable for take off directly from Earth)? If we are establishing human settlements on the Moon (with efficient and economical transportation from Earth) practically every kind of astronomical observatory that is presently done in Earth orbit or on Earth itself may be performed more efficiently on the Moon. Indeed, precision pointing of telescopes, important engineering problems for satellite observatories, will disappear when we can point telescopes working against the virtually infinite (when compared to the infinitesi- mal masses of the telescopes) mass of the Moon. In this article, we will primarily consider case (A), in which the Moon provides unique observational opportunities not readily available elsewhere in Earth's vicinity.

[II]

There are several advantages to be achieved by placing telescopes on the Moon, i.e., (a) expanded spectral coverages in the vacuum of the Moon, (b) improved resolving powers for telescope by the absence of the blurring terrestrial atmospheres, (c) the Moon as an immense optical bench for interferometry, and (d) the radio-noise-free far side environment of the Moon. (a) Astronomical frontier has expanded enormously over the past century by increasing the resolving power of telescopes and expanding the spectral range, which had largely been limited to 1 the visible light throughout much of the history. The vacuum environment of the Moon removes all obstacles due to terrestrial atmospheric absorption and blurring. Placing observatories in Earth orbit can also remove the atmospheric absorption and expand the spectral coverage. In the near future, before human bases on the Moon will make it practi cal to build there all sorts of observatories at costs competitive to those for observatories in Earth orbit, orbital observatories will continue to serve as vital tools for astronomical research. (b) The resolving power of telescopes has been expanded by making ever larger telescopes.

When the turbulence of the Earth atmosphere presented an obstacle in achieving the diffraction limit of large mirrors, we have substantially overcome the obstacle by designing adaptive optics that can circumvent the problem. New mirror making technologies have enabled us to build huge telescopes that are 8 to 10 meters across, as compared with the 5-meter telescope of Palomar Mountain that dominated astronomy for several decades. It must be noted here that even using adaptive optics, we need to go outside the turbulent, blurring terrestrial atmospheres to achieve the ultimate theoretical diffraction limit of a telescope. (c) Interferometry using the Moon as an immense optical bench in vacuum: There is another branch of astronomical observations that are finally coming of age. That is astronomical. interferometry. In astronomical interferometry, we are pushing astronomical resolutions way beyond what is possible with large optical telescopes; indeed, the separation of two telescopes becomes effectively the same as the aperture of a much larger telescope in the direction of the separation of the two telescopes. If you have several telescopes in an interferometric array, we will have that high resolution in all those directions connecting the telescopes involved. This technology is already being tested and proven effective at Mount Wilson.

However, the atmospheric blurring still presents limitations on the ground even with the use of adaptive optics. The terrestrial planet finder program envisions, among other possibilities, several telescopes mounted on a solid beam in space, which is perhaps a few tens of meters long. For pratical reasons, that is about the maximum length currently conceivable for the interferometric beam in orbit. The high-resolution is achieved in the direction across the beam. There are also plans for using several electronicallylinked, free-orbiting telescopes. This is an attractive concept that should be tested and explored. However, the distances between and the relative directions of the individual telescopes must always be known with extreme precision. We are talking about the accuracy in the range of angstroms. [one angstrom is about the size of an atom.]

There are many technological problems that must be overcome before the concept of interferometrically-linked orbiting telescopes becomes a feasible idea. For astronomical interferometry, the Moon provides a large solid base upto a few thousand kilometers across. The separation between the telescopes can be in the range of kilometers or even tens of kilometers rather than a few tens of meters in Planetary 2 Finder type missions (which would nevertheless make great advances in astronomy). For this purpose, we might think of the Moon as an immense optical bench in vacuum. An array of interferometrically linked telescopes on the Moon does not require an unknown technology not yet tested on the ground. (d) Radio observatories on the far side of the Moon: The only exception to the restrictions placed on the resolving power of ground-based telescopes has been in the radio wavelengths.

Throughout much of the radio wavelengths, observations can be made from the ground using radio telescopes without the interference of atmospheric absorption and blurring; radio observations, which was began by Carl Jansky in the late 1930s, began flourishing in the 1950s. However, there is much radio noise in the Earth environment; it is difficult to shield radio telescopes from all radio noises no matter where we locate radio telescopes, even in mountain craters as we do in Puerto Rico. There is one place where radio noise can virtually be eliminated. That is the far side of the Moon; the solid rock of the Moon itself will shield radio observatories on the far side (the side of the Moon that always faces the direction away from Earth) from radio noise generated on Earth.

[III]

Exploratory robotic missions to establish remotely controlled observatories on the Moon to test the Moon sites for astronomy: Before a manned Moon base is established to enable building and operating of lunar observatories, it is possible to test their feasibilities first at reasonable costs. In one such plan, an array of two to three interferometric telescopes, mounted on rovers, may be landed on the Moon and placed in preplanned locations. A suitable rover has been fieldtested in an Antarctic crater and could be modified for a lunar mission. Such an experiment can be conducted at costs comparable to that of Explorer class observatories in Earth orbit. Other types of robotic telescopes could also be placed on the Moon to test the practicality of the Moon for astronomical observatories. This could be done, if for no other reason than to demonstrate that "moon quakes" or dust kicked up meteoric impacts are not problems for astronomical observations. Some scientists express concerns even though there is no evidence that possible problems exist. These forerunner prototype observatory missions could put these concerns to rest, once for all. Ad astra!

References: A few recent papers on how to set up low-cost astronomical observatories on the Moon are listed below. The proceedings of an international symposium on. observatories in Earth orbit and beyond is also listed at the end. Kondo, Y., Terzian, Y., Chen, P., Mendell, W.W., and Oliversen, R. "You Can See Forever from the Moon", Artemis, 2, 14, 2000. Chen, P., Kondo, Y., and Oliversen, R. "A Realistic Attainable Lunar Astronomy Mission", in 'Highlights of Astronomy', Kluwer Academic Publishers, 1997. Chen, P., Kondo, Y., and Oliversen, R. "Moon as an Immense Optical Bench in Space", IAU JD No. 23 'Astronomy from the Moon', 1997. Chen, P., Oliversen, R., and Kondo, Y. "A First Generation Lunar UV Observatory", Proc. 'Space 2000', ed. S.W. Johnson, K.M. Chua, R. Galloway, & P. Richter, Amer. Soc. Civil Eng., 712, 2000. Kondo, Y. (editor) "Observatories in Earth Orbit and Beyond", Proceedings of the International Astronomical Union Colloquium, Kluwer Academic Publishers (1990).

In the 1980's Frank White helped President Reagan's National Commission on Space prepare its report, which laid out a vision of space exploration and settlement that was the most profound and far reaching since John Kennedy's famous speech committing us to go to the Moon. His original book "The Overview Effect" is one of the important works of the Frontier movement. A gentle and quiet man, Frank speaks softly, but his words are profound. Just what will happen when the first generations of humanity are born "out there" and look back from their homes on the Moon and Mars and the asteroids to that "other" world known as Earth? Such arguments as this and the other more spiritual elements of this collection may not be as immediately obvious to the reader as the scientific, strategic, technological and economic arguments for us to return to the Moon, but they are in actuality the Real reasons we must do so.

The Overview Effect from the Moon

by Frank White

As we look back on the early days of space exploration, we can clearly see that, in the words of Apollo astronaut Gene Cernan, there were "two space programs," the one that took place in Low Earth Orbit (LEO) and the lunar space program that sent 24 people to the Moon[1].

Both programs produced a phenomenon that I have labeled "The Overview Effect." This experience, on its simplest level, is triggered by seeing the Earth from space and realizing the planet's inherent unity and interconnectedness. It is an experience that I have documented by interviewing people who have traveled into orbit or to the Moon. It varies with the experiencer, but always seems to be present in some form. Later developments have shown the Overview Effect to be more than a momentary experience of individual space travelers, however. What we now call "globalization" is in fact a manifestation of the effect. It flows from the "technological Overview Effect" produced by satellites that transmit information around the globe in a near-instantaneous manner. We are today living in a planetary civilization that has resulted from the permanent establishment of the technological Overview Effect.[2]

With the recent discussion of the case for returning to the Moon as part of the President's new space vision, one reason that has not been mentioned very often is still highly relevant: the impact on human consciousness of the view of the Earth from the Moon. To be specific, this is the "overview" that should be taken into account. The view of the Earth from orbit is by no means the same as the view from the Moon. From orbit, the Earth is still quite close, 100 to 200 miles away. You cannot see the entire planet from that vantage point, only a part of it. Standing on the Moon, the space voyager is 240,000 miles from the surface of the Earth, and the planet is seen for what it is: a jewel-like orb suspended in black space, a tiny oasis for humanity in an enormous and largely unknown universe. It is for this reason that the Apollo 8 mission to the Moon, which took place during the Christmas season in 1968, was so profound. For the first time, astronauts were far enough away to turn cameras back and show us the whole Earth, which they did. It was the first time that the people of the planet fully experienced the Overview Effect. While this moment is history has been somewhat overshadowed by the Apollo 11 Moon landing less than a year later, Apollo 8 was in many ways just as important in terms of changing human consciousness.

There are many other differences between the orbital and lunar views of Earth. From the Moon, the Earth has phases, like the Moon does when it is seen from the Earth. When humans stand on the Moon, it seems that the relationship of satellite and planet have been reversed.

The number of trips by humans into orbit now numbers in the hundreds, while the number of journeys to the Moon remains small. Moreover, flights to orbit continue to this day, and astronauts permanently occupy the International Space Station. However, Gene Cernan was the last person to walk on the Moon, and he returned to Earth in 1972, more than three decades ago. For this reason, we have relatively little data on which to base any conclusions regarding the long-term impact of large numbers of people seeing the Earth from the Moon over an extended time period.

While the Overview Effect has had an enormous impact on human thought and human society, what we have seen so far is really only the beginning. A permanent lunar presence would carry forward the revolution that the early space program began. More and more humans would have the direct experience of seeing the Earth as a small planet suspended in space, and an even greater number would have the indirect experience. Some day, the first child to be born on the Moon will look back and see the Earth, and the Overview Effect will be a common component of her life, and that of all humans born off the home planet thereafter.

In addition, we should keep in mind the point made by Edgar Mitchell, the sixth person to walk on the Moon. When I interviewed him for my book, he suggested that the impact of the space flight experience depended on the experiencer's willingness to be open to and affected by the experience. That being said, he characterized the view from the Moon as "it gets you closer to a more universal experience because of the distance and wider view. You identify more with the universe as it is instead of the Earth as it is."[3] Here, Mitchell is talking about something that is a normal extension of the Overview Effect, which is "the Universal Insight"—a realization that we are as much a part of the universe as we are of the Earth.[4] If we take Mitchell's two points together, the case for the Moon strengthens in terms of its impact on consciousness, especially for those who might be *seeking* shifts in awareness, or spiritual awakening. Imagine a retreat center on the Moon, or a monastery. Such a place would be visited, or inhabited by, people who had gone there with the intention of being open to the experience and who could well have a powerful result, given the distance from Earth and the wider view of the universe itself. One might even choose to establish such an institution on the side of the Moon permanently facing *away* from the Earth, thereby flooding the senses with the Universal Insight and moving far beyond the Overview Effect. Or consider a place where government leaders might come together for summit conferences (this common term would clearly take on new meaning.) In the words of Apollo astronaut Michael Collins, "I really believe that if the political leaders of the world could see their planet from a distance of…100,000 miles, their outlook would be fundamentally changed."[5] Perhaps not all the pressing problems of the Earth would be solved by such an experience, but it seems hard to imagine that participants would make the same kinds of decisions that they have made for centuries in Paris, Rome, Geneva, and Washington. The lunar version of the Overview Effect was anticipated in a short story written in the 60s by Thedore L. Thomas called "The Far Look." He described the eyes of astronauts returning from the Moon as different: "There was a look in those eyes of things seen from deep inside, of things seen beyond the range of normal vision. It was a far look, a compelling look, a powerful look…"[6]

While we know more than Thomas about the far look, there is so much more to learn that it's somewhat puzzling how long we have avoided finding out by returning to the Moon. We can say that wherever human beings reside in space, they bring, through the symbiosis of their human systems with natural systems, a new kind of system, an "overview system" that becomes a locus of consciousness in the universe. As people begin to live permanently on the Moon, they will create a uniquely lunar overview system that will be an important step in human evolution. In fact, *The Overview Effect* concludes by calling for the creation of a "Human Space Program," a planetary commitment to explore the universe over the next millennium. Its declared purpose is "to support humanity's understanding and achievement of its purpose as an active partner in universal evolution, creating overview systems that increase conscious awareness throughout the universe."[7] It is my hope that creating the Human Space Program will become an important component of the case for our return to the Moon.

We cannot know where this will lead, only that it will be another step in the long evolutionary journey of human beings as we become true "Citizens of the Universe." This is a step we must take, and it is a step that we will take. The only questions are, how soon will it happen and how bold will we be in making it happen?

[1]Frank White, *The Overview Effect: Space Exploration and Human Evolution*, AIAA, 1998, p. 183.
[2]Ibid, p. 55.
[3]Ibid., p. 203.
[4]Ibid., p. 34.
[5]Ibid., p. 187 (from Collins's book *Carrying the Fire*, Farrar, Straus, and Giroux, New York, 1974.
[6]Thomas, Theodore L., "The Far Look" in *Spectrum 5,* Kingsley Amis and Robert Conquest (ed), Berkley, New York, 1966. I am indebted to Gary Oleson for re-introducing me to this classic science fiction tale.
[7]*The Overview Effect*, p. 171.

Frank White Born in Greenwood, Mississippi, Frank has long been active in the space exploration movement. A member of the Harvard College Class of 1966, he concentrated in Social Studies and graduated magna cum laude. He was elected to Phi Beta Kappa, and attended Oxford University on a Rhodes Scholarship, earning an MPhil in 1969. He is the author or co-author of six books on space exploration and the future, including The Overview Effect, The SETI Factor, Think About Space (with Isaac Asimov), March of the Millennia (with Isaac Asimov), The Ice Chronicles (with Paul Mayewski), and Decision: Earth. Frank is married to Donna White, and he and Donna have five children and three grandchildren.

At the beginning of this book, I spoke of such things as destiny and hope. I tried to contextualize our return to the Moon in a much bigger and ongoing activity—the expansion of life and humanity beyond the Earth. Each of the essays herein presented in some way play into that larger context. Some are very specialized, some very broad in scope. Yet each is spoken from a place in the writer that grasps the import of this cause. Steve Wolfe reaches into that place for us in this essay and exposes the almost metaphysical rationale for what we are about to do. And yet, Steve is not just some ivory tower philosopher. Like many of those in the Frontier movement, Steve is about achieving results. He is about actions, not just words. To this end, while a staffer for the late Congressman George Brown of California, Steve crafted and helped his boss introduce the Space Settlement Act of 1988. As odd as it sounds, until this law was passed, it was not the goal of the United States to open the frontier to its citizens. Steve changed that, at least for a while. (After Steve left Washington, the law was killed as part of the paperwork reduction movement– a convenient thing for the space agency at the time, which had no interest in following the bill's mandate that it report progress on space settlement to Congress on a regular basis!) The core points Steve makes in this piece are what gets people like he and I up in the morning. Name it a "calling" or a "drive" or what you will...

The Conscious Evolutionary Choice

by Steven Wolfe

Sunday, July 20, 2025

Here's the birthday gift that came into my mind this morning
as I woke up—just two lines:

The Destiny of Earthseed Is to take root among the stars.

I don't know how it will happen or when it will happen.
There's so much to do before it can even begin. I guess that's to be expected.
There's always a lot to do before you get to go to heaven.

Octavia E. Butler, from Parable of the Sower

The reason we go to the Moon is the same reason we go anywhere in space: to build a knowledge base that will assist in the creation of permanent communities beyond Earth.

Humankind engages in space activities for no other reason than to fulfill a primordial demand to do so. That demand is an *obligation* we have to the planet that gave us birth—an *obligation* to carry the seeds of life to other celestial shores. Why? Simply so the life that evolved on this planet can be assured that it is ongoing, regardless of what happens to this particular host world. All of the technology, entrepreneurship and politics behind this great endeavor does not amount to anything more or less than that.

Though returning to the Moon for the sake of fulfilling *human destiny* is reason enough, many in the space arena do not see a place for such declarations in serious space policy discussion. Regrettably, this attitude has served to seriously inhibit the progress of the human space program.

The challenge of this essay is to create some room where we can view space development in the broader context suggested above. If we can create that room, just maybe we will see how correct and appropriate returning to the Moon truly is. This, however, requires a willingness to see our efforts in evolutionary terms that extend far beyond any programmatic timeline. If we can take this longer view, we may be able to grasp just how essential human expansion into space is, as well as demonstrate why lunar development is the correct focal point for the immediate future.

Seeing the space program in evolutionary terms has little significance unless we are willing to *consciously* take part in the evolutionary process itself. As I will explain, our civilization is, for the most part, *unconscious* to this process. This is not surprising as humanity is only just beginning to understand its own capacity to *consciously evolve*. The reason we should care about this distinction at all is because our very success in becoming a multi-planetary species may actually depend on our participation in the evolutionary process *consciously* and deliberately.

From a *consciously evolved* perspective, lunar operation, as an initial focus of the global space program, becomes the glaringly obvious choice.

But before we explore the *rightness* of going to the Moon, let's get more clarity around just what it means to *consciously evolve* into space.

Conscious Evolution

As we begin the 21st century, we generally accept that humanity is the product of millions of years of evolution. There is also a growing awareness that the biological evolution on Earth is just one manifestation of the process of evolution that has been on-going since the "Big Bang." As Eric Chaisson wrote in *Cosmic Evolution,* "we have discerned a common basis on which to compare all material structures, from the early universe to the present—again, from Big Bang to humankind inclusively."

What is less appreciated is that humanity, in its current evolved condition, is imposing enormous influence on the continuing evolutionary course—whether we choose to be *conscious* of it or not. In *Conscious Evolution,* Barbara Marx Hubbard wrote,

"We are an integral part of the evolutionary journey. In our genes are all generations of experience. In our genius is the code of conscious evolution. In our awakening lies the patterns of the planetary transition from our current phase to the next phase. Our mind is designed to know the design of evolution toward higher consciousness and freedom."

Frank White in *The Overview Effect* wrote, "Humanity…has the singular opportunity to guide and shape its own evolution, working in conscious partnership with the whole."

Thought leaders in the field of consciousness and spirituality have been stressing the critical importance at this juncture for humanity to move from *unconsciously* influencing evolution—which is having a disastrous impact on the planet—to being *consciously* engaged in the evolutionary process. The spiritual teacher, Andrew Cohen said,

"When you awaken and suddenly recognize your own place in the evolutionary process, you realize something BIG: that it's all up to you…If the evolutionary potential inherent in consciousness is going to be activated, you have to be responsible for it. Why? Because it is only through the human vehicle that the creative principle, the God impulse that initiated this whole process, has the capacity to know itself."

Waking up to the demand to *consciously evolve* has a different implication for each individual who is sensitive to the call. In this essay, I am speaking specifically about the *God impulse* to evolve beyond this planet. This response to the evolutionary impulse, however, is just one manifestation. This very same impulse is calling people to action in many critical areas, including global conflict resolution, environmental activism, human rights, hunger, and myriad other noble works. We are not likely to succeed in any of these areas, however, unless we can *consciously* engage in the evolutionary process.

Imagine a pilot who wakes up in the cockpit of a plane flying on autopilot. The flight log tells him the plane is heading toward a particular destination—but he cannot possibly arrive safely without manual assistance. Awake to his situation, the pilot can improve his chances, by determining the speed and course, as well as avoiding unexpected obstacles. If he had remained asleep, the plane would have kept going for some distance, but eventually it would run out of fuel, and crash, killing everyone on board.

The space arena is much like that sleeping pilot. We are, for the most part 'asleep' to the evolutionary context in which expansion into space is unfolding. We like to think of our space program as very directed and purposeful, and it is, to a point. But if we step back over the hundreds or even thousands of years of mankind's activities, we see a much different picture than the one that exists through the narrow lens of the first decade of the 21st Century. Space projects today are rationalized and measured in very limited, near term objectives, even those with a 20-year planning horizon. We are asleep to the fact that space migration has been our destiny all along, and that our engagement in space development activity is occurring *regardless* of how we narrowly choose to rationalize these actions within our present cultural framework. To the point, the awakened view of conscious evolution recognizes that the space migration imperative existed *first*, and that the mind's social or political rationalizations for going into space came *second*.

Fortunately, for the moment, our *unconscious* evolutionary processes—our autopilot—is still functioning. The Bush Administration made the correct decision by choosing to establish a lunar base as the next phase of the U.S. manned space program. The White House made the correct choice, despite its inability to place these near term steps in the context of human evolution.

Another good metaphor here is the ant swarm creating an anthill. Ask a single ant what it is doing, and in what part and context it is involved. It will have no idea. At best, it might say it's carrying a grain of sand from here to there. Like ants, the majority of participants in the space community have a viewpoint that encompasses only a fraction of the whole picture, and therefore few involved have any sense of the full reality they are in the process of creating. Even the Moon-Mars initiative, as visionary and pivotal as it appears, is still a small swatch on an infinitely bigger evolutionary tapestry.

Some might falsely conclude that the "autopilot" of evolution will eventually carry us *unconsciously* to our solar civilization in some vague and distant future, so why worry about it now? To take such an attitude is to have little appreciation for the enormous gulf that exists between the President's unveiling of the Moon-Mars program and the actual construction of a lunar base—never mind the gulf between a lunar base and a human settlement on the Moon. The odds are massively stacked against success for NASA in this regard. The history of our human space program since the days of Apollo is painful proof that our persistent *unconscious* march into space is simply not enough to spark the age of space migration.

Of course, we are not ants. We indeed have the capacity to hold in our minds the biggest possible context for our existence, including space migration, if we are only willing to open ourselves to that context.

It is not enough, however, to grasp the meaning of human evolution into space *intellectually.* As was mentioned, this is a primordial call we are responding to. Therefore our further challenge is to sense that calling, *experientially at the deepest part of our being.* By this we are seeking an emotional connection to the very source of that calling, that can only be made through deep contemplation and meditation. The depth of understanding being discussed here can't be achieved through intellectual discourse alone.

Once we establish a clear position of consciously engaging in the evolutionary process, there is the potential for several things to occur within us. First, we become acutely aware of where we are personally with respect to the ultimate fulfillment of the space settlement imperative. Second, we realize that we actually have within us the capacity to reach the desired outcome (both individually and collectively). Third, we become sensitive to internal and external messages about the appropriate actions we should be taking consistent with this calling. And finally, we begin to care less about our own desires or that of our particular *fiefdom*, and give ourselves more fully to the purpose to which we are being called, for the benefit of all. We are uniquely bonded with every one else pursuing this dream, in a way that transcends our personal interests or place of work. It is no longer about getting our piece of the space pie; it is about how we can best serve that which is calling us into action.

As Martin Buber wrote in *I and Thou,* "the free man …intervenes no more, but at the same time, he does not let things merely happen. He listens to what is emerging from himself, to the course of being in the world, not in order to be supported by it, but in order bring it to reality as it desires."

This new orientation shapes and informs every decision we make in the near term. This knowledge puts our efforts into a much greater context from which we can draw meaning and inspiration to help sustain us through difficult periods. While we may derive benefits from space activities in the near term, and we should expect to do so, we fundamentally know that our efforts are part of something much larger and much more important.

With this perspective we also realize that "it is all up to us," as Cohen said, if we are going to succeed in building solar civilization. To offer one more metaphor, we are like the chick that must, of its own determination, free itself from the eggshell or it will not survive. On some basic level, the chick must wake up to its condition in its shell and realize that it must break out or die. It can't receive help from anyone. If it does, then it will not likely survive because its muscles will not properly develop as they do in order to break through the shell on its own. Like the chick, we must wake up to our situation and earnestly take hold of our destiny. God will continue to assist us in our evolution, but He will no longer do it for us. He's already given us everything we need.

The challenge for those who feel the calling of space is whether they can recognize this predisposition as an *obligation* that was cast for humanity at the very dawn of its existence—regardless if they call it God's will or cosmic DNA encoding or something else. Are we willing to consider and ultimately embrace humanity's purpose as the carrier of life beyond this planet? For those who can move to that position, the responsibility before them becomes much more intense and urgent.

The conscious evolutionary stance is not one that can be achieved through casual consideration. It must be sincerely cultivated through deep contemplation of and meditation on everything we think we know about the human purpose in space and our own motivation in that regard. This is a process that can't be rushed, and if we are serious, never really ends.

I am of the opinion that only when enough of us can embrace this perspective will we ever be able to make real strides into space.

The Moon as the Consciously Evolved Choice

From the perspective of *conscious evolution*, a lunar research facility is in one sense a minor consideration. The decision to establish the Moon as the logical starting point for human exploration and development of the solar system seems so obvious as not to be worthy of discussion.

Unfortunately, the space political culture is dominated by an "all or nothing" mentality that poisons every decision. We have only so many space dollars, the limiting rationale goes, so we are forced into a competition to pick the one best way to spend them. In this environment, it is tempting to latch onto scenarios that promise to 'leap frog' the Moon and go to straight Mars, or a large-scale orbital city, or something else. Tragically, this cultural psychology of limitations has polarized those people who should be working together in their shared sense of purpose in opening space for all humankind.

Through the lens of conscious evolution, concerns about programmatic budgeting are seen merely as part of the process of reaching the desired goals in space. Yes, we must properly organize our financial resources to meet the task, and that will undoubtedly be very difficult, as will many other tasks involved in extending civilization into space. But, fundamentally we view these obstacles as merely issues to be worked through as we move forward. There is no longer a question of whether or not we can afford such projects, any more than there is a question about whether or not we should put roofs over our head or food on our tables. We simply recognize that this endeavor must be supported, and proceed accordingly. In this frame of mind, we are also less reactionary to the immediate demands, especially those imposed for self-serving purposes.

Acting from a conscious evolutionary position, we will experience a flow, personally and collectively, as we align our resources for greater efficiency and creativity in support of the space mission. This alignment by definition will include the function and interrelationships of existing institutions responsible for making the decisions that drive the course and pace of our march to space, which include government agencies, private industry, academia, and non-profit organizations. Like the ants constructing their hill, we are smoothly in tune with creating the future into which we are consciously evolving.

To understand the unique importance of the Moon, however, we must understand that to *consciously evolve* is to also align with the natural ebb and flow of universal forces. In this context, space migration has an organic quality. If we look to nature for examples, *proximity* is key to biological expansion. Organisms will seek to establish themselves on the closest available territory that can sustain them.

While it's true that the errant seed can catch an upward draft and be carried to a distant spot, the dominant pattern is to spread locally. Some have set their sight on the Moon and others on Mars. Still others long for the free flying spheres envisioned by Dr. Gerard O'Neill. This is no different from the reach of a vine: one branch will crawl up a wall while another chooses a fence, both destinations are equally appropriate. The vine only cares to reach outward.

Though the Moon may yet prove itself to be an untenable host, we are naturally inclined to seek first to establish ourselves there before moving out further. We focus our attention on the Moon at this point, not because it is any better than any other place in the solar system, but because it is the most easily accessed.

It is important to mention that in the course of using the Moon as a test bed, we will possibly see the emergence on the Moon of the first self-sustaining habitat beyond Earth—the first offspring of Gaia, husbanded by the human race. The first-born will ever be in easy reach of its mother. This is not an insignificant psychological plus to stepping off this planet to the Moon.

As I near the end of this essay, I will risk offering a seeming contradiction to what has come before. Although the Moon is the obvious, or organic, choice for the first space settlement, from a *consciously evolved* perspective, it also does not matter whether it is or not. The entire solar system is our domain, after all. In a consciously evolving context, there is an *ease* with which we take our actions. We are more interested in *allowing* the future in space to unfold "as it desires," rather than insisting that it occur in any particular way. This may seem a contradiction but it is the only authentic way to be in relation with conscious evolution.

The Moon holds a profound position in the story of life's emergence into the solar system regardless of how we proceed. In all likelihood, she will serve as host to Gaia's first offspring. But, if she hosts the second or third child, no matter. For Gaia will, like all good mothers, love all her children equally.

Steven Wolfe spent five years as the legislative aide for space policy to the late Congressman George E. Brown, Jr. (D-CA), where he drafted the Space Settlement Act of 1988 (contained in Public Law 100-685). Steven served as Executive Director of the Congressional Space Caucus and on the Board of the National Space Society. He has appeared on radio and television commenting on space issues. In the 1990's, he built a career in the management of not-for-profit organizations, and currently serves as an executive for a major trade association based in New York City. Recently, Steven helped William E. Burrows and Robert Shapiro develop their idea for a lunar archive project called the Alliance to Rescue Civilizations (ARC). For the past three years he has been engaged in an inquiry into the relationship between space development and spiritual enlightenment.

Throughout this book, there is an underlying tone of the Moon as Frontier, as the next place we will be going to live beyond the Earth. As such, it presents us with the chance, and challenge , to explore not just its physical realm, but also to explore the systems and society that will best make of it a vibrant and ever growing extension of the human domain. Ed Hudgins, a diehard disciple and one of the leaders of the movement started by Ayn Rand, believes it can be a place where we can develop her ideas of individualism. It is an exciting concept, and as others in this anthology have pointed out, the choices we make now will determine whether the soil of the Moon becomes the fertile bed in which the seeds of a new culture are born, or the grave of our last brave attempt to reach beyond ourselves.

The Social and Spiritual Significance of Lunar Settlement

by Edward Hudgins, Ph.D.

On July 20th, 1969—the day Neil Armstrong and Buzz Aldrin became the first humans to walk on the Lunar surface—it took a leap of imagination to believe that after only five more visits, nearly four decades would pass with no new footprints on that surface. The problems were economic and political; support from elected officials for the costly program and thus taxpayer dollars dried up.

President George W. Bush's announcement that he wants NASA to aim again at Earth's natural satellite, this time to establish a permanent base, was greeted with cheers by space enthusiasts, but with skepticism by elected officials who see another boondoggle with escalating costs. These concerns are legitimate, and that is why free market-oriented space advocates understand that a return to the Moon should have significantly more private sector involvement.

For example, former Rep. Bob Walker (R-PA), who was a member of the President's Commission for Implementation of U.S. Space Exploration Policy, suggested that the federal government might simply offer a 25 year exemption from taxation for any company that can build a full time Lunar facility. Companies like General Electric and Microsoft, imagining decades free from taxes, might turn their imaginations and billions of dollars to the task of producing such a base. This approach would not require any outlays of tax money. And, even though the government would forego tax revenue from the winning company after the base was built, it would gain considerable revenues from all of the economic activities involved in the development and construction of a Moon base, to say nothing of the new technologies and infrastructure that would result.

But these and other private sector approaches for a return to the Moon have profound implications not only for the development of technologies and infrastructure, but also for the nature of the space-faring civilization that might result.

Settlements and Civilization

A major motivation for those who long to settle the Moon, Mars and other worlds is the same as for those who settled America: the opportunity to create new, better societies. Pioneers came to the shores of the New World because they held the value of their own lives and happiness so high that they would accept only the best they could achieve. They realized that they had to take the initiative, to use their wits to figure out how to cross oceans and continents. They realized that risks were part of the bargain, but that the prize of happiness was so great, that such risks, literally of their lives, were worth taking.

Most Americans shared a particular vision of a good society, that is, one in which they as individuals would be free to live their own lives as they saw fit, to raise their own families, farm their own farms, run their own businesses, to worship as they please or not at all. Their aims were economic, political and ethical. But most important, they were personal and they did not involve infringing on the rights of others.

The credo of this country was best expressed by Thomas Jefferson in the Declaration of Independence, that we are endowed with "certain unalienable Rights, that among these are Life, Liberty, and the Pursuit of Happiness—That to secure these Rights, Governments are instituted among Men, deriving their just Powers from the Consent of the Governed."

The result of this philosophy of individualism was the freest, most prosperous and creative country the world has ever known—a country that sent men to the Moon.

Science fiction has been the arena in which new settlements on other worlds have already been explored. One of the most notable examples is found in Robert Heinlein's *The Moon is a Harsh Mistress*. In this story about a Lunar war of independence against the Earth, one computer "character" that leads the revolt is described as the "John Galt" of the revolution after a mysterious, hidden, character who leads a strike of the men of the mind in Ayn Rand's epic novel *Atlas Shrugged.*

Rand's vision is, in fact, the one that should inform real-life Lunar pioneers and settlers, who wish to make the Moon part of humanity's domain and to add a new chapter to humanity's story as profound as the Renaissance and Enlightenment. This vision not only will facilitate such a civilization but will be indispensable—given the nature of settlers on the Moon and the challenges they will face—if a Lunar colony is to develop into such a civilization.

Who Pays to Settle the Moon?

Consider first the economic as well as cultural aspects involved in how Lunar settlement is financed and, as a result, the type of civilization that emerges. If the governments foot most of the bill, costs will of necessity be high because this is always the case for government-provided goods and services.

For example, when the American space station was proposed in the mid-1980s, it was supposed to cost $8 billion, accommodate a twelve-person crew and be in orbit by the 1990s. But the station was redesigned many times, its size reduced and international partners brought into the project. By 1995, one report by the U.S. government's General Accounting Office found that, through June 2002, the actual cost of designing, building, and launching the station would be $48.2 billion. The cost of operating the station after its assembly through 2012 will add another $45.7 billion to the price tag for a total bill of $93.9 billion. New estimates put the final cost of the station even higher.

Whether it's Amtrak providing transportation or the U.S. Postal Service delivering the mail, government provision of goods and services means higher costs, taxpayer bailouts and problematic quality. By contrast, private providers over time reduce costs and improve quality in order to market goods and services to paying customers. For example, from the time airlines were deregulated in 1978 through the year 2000, the real cost of flying dropped by at least one-third; Americans in 1978 made 275 million trips on commercial airlines while in 2000 they made 650 million trips. Whether the products are automobiles, consumer electronics or computers, whether the service is airline flights or the Internet, the private sector brings down the costs.

If the return to the Moon is principally government-led, the base there will likely have all the cost and quality problems of the International Space Station. The facility will be expensive, always struggling and will never grow into shining Lunar cities.

If governments remain in charge of a Lunar base under the Outer Space Treaty, or, worse, the Moon Treaty, the Moon will always resemble Antarctica under the 1958 international agreement that barred private property and commercial activities: a frozen, dead wasteland.

However, if private parties rather than government bureaucrats lead the way, the results will be far more favorable. Entrepreneurs and property owners will want to do what brings them the best results. Entrepreneurs, like scientists, experiment to find the best strategies and move quickly, unlike government bureaucrats who are forever doing studies and reviews, seeking approvals and permissions, and maneuvering among ever-changing political demands and mandates.

Imagine the different culture and civilization that would emerge on a Moon settled by private creative innovators seeking tax freedom, as suggested by Rep. Walker's plan, versus those that would emerge from government functionaries.

Private compacts

Ideally, even if governments foot part or much of the initial bill for returning to the Moon, only private settlement will lead to a Lunar civilization. Ideally, private parties will form consortia to establish such settlements. It is crucial that these consortia draw up sound principles of operation between members before departing.

In American history we see, for example, the Mayflower Compact, in which the Pilgrims agreed to a form of self-government *before* leaving for America. Settlers crossing the continent usually would make contracts concerning who owed what services to whom and how the members of the group would govern themselves.

But some agreements are good and others are not. The Jamestown colony, founded in 1607, had a bad compact with bad results. The company that sponsored the colony made provisions for settlers to be fed from a common store. There was no incentive to be productive. But communism did not work. Gentlemen settlers spent their time hunting for gold—they found none. After less than a year, of the 104 original settlers only 46 were left alive. John Smith later instituted a new rule: Those who do not work shall not eat. A revised compact gave settlers an incentive to produce food and other products, and pay the sponsoring company from those proceeds.

Economic institutions create incentives that form the moral habits of individuals. A socialist system that treats adults as children who can't tie their shoes or wipe their noses produces sniveling subjects; a free market system produces proud, productive, creative citizens.

Those who create consortia must keep these facts in mind. Any compact for Lunar colonists likely will initially require all individuals to work together to produce common necessities that all will need to consume—air, water, food. Those who create the consortia should resist the temptation to keep individuals bound into such a system. The best compacts will have easy opt-out provisions.

For example, if certain individuals in the consortia can figure out a way to produce oxy-

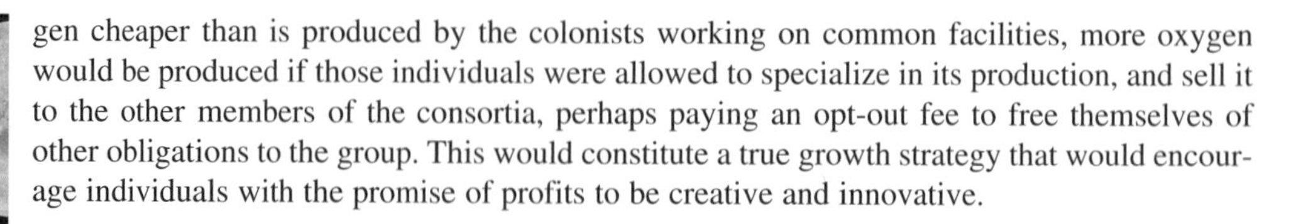

gen cheaper than is produced by the colonists working on common facilities, more oxygen would be produced if those individuals were allowed to specialize in its production, and sell it to the other members of the consortia, perhaps paying an opt-out fee to free themselves of other obligations to the group. This would constitute a true growth strategy that would encourage individuals with the promise of profits to be creative and innovative.

Spontaneous Order

Lunar pioneers and entrepreneurs will need unique skills to survive and flourish. Nearly all will need to possess scientific, engineering and technical knowledge. But those with such skills often make a mistake when viewing society. They tend to think that just as technocrats can figure out what kind of equipment might be needed to perform a certain task, then design and build it, so societies can be planned and engineered.

But the lesson of every socialist or welfare state plan is that this is not possible. The success of the free market results from each individual with their own unique knowledge and perspectives experimenting, offering the goods and services they think will sell. Most entrepreneurs meet failures, learn from them and go on to succeed. When government planners fail, they usually raise taxes to subsidize their failures which usually command the allegiance of strong political constituencies. Good money is thrown after bad.

The dynamics of how institutions and wealth emerge from human activity, but not human planning, can best be understood with reference to the insights of economist and social thinker F.A. Hayek. Most people conceive of order in the world as falling into one of two categories. First trees, mountains, and solar systems arise and evolve naturally. Second, watches and tables, statues and rockets result from intentional human planning and action. This latter conception gives rise to the belief among socialists and statists that all wise, caring bureaucrats can plan and benevolently guide economies to prosperity, and among social conservatives, that censors and vice squads can create civil societies. But Hayek identifies a third type of order, spontaneous order that arises from human action but is not specifically planned by men. A classic example of this type of order is money. In primitive societies, individuals might need to travel long distances to trade with one another. But it is difficult and costly to tote four cows, seven bales of hay, and a slab of copper over a mountain, all to be exchanged for a dozen sheep, six large jars of barley, and a handful of gold nuggets. Many merchants hit on the strategy of trading their goods simply for those small yellow pieces of metal rather than sheep, which tend to wander off, and the rest of it. Gold is easy to transport, difficult to counterfeit, durable, rust-resistant, and easily divisible. Traders can take the metal home to exchange for what they really need. Thus as individuals sought to exchange goods more efficiently, the institution of money emerged and became an economic institution as an unintended consequence. No one invented it.

In a new society on the Moon, the dynamics understood by Hayek will be even more essential than on the Earth, because on the Moon there will be less room for errors, and greater need for institutions that can change and evolve quickly to meet the needs of pioneers and settlers.

A Lunar Credo: Reason, Responsibility, Justice and Benevolence

Those starting new societies bring with them beliefs and habits that will help determine whether they succeed or fail. Lunar settlement will offer the opportunity for individuals to bring with them only the best beliefs and habits appropriate for a bright future, leaving the ones that have weighed down humanity on the heavy lands of Earth. The following four particular values, with corresponding virtues and habits, will lead the way to this future.

Reason—The first is a commitment to reason. Human history has been plagued with every form of irrationality, religion and superstition that has brought repression, misery and deaths by the millions, whether in the name of creeds about heaven, race or class. Even today, in our advanced society that is the result of the Enlightenment philosophy which provided the underpinnings of America, too many individuals do not know the difference between science and scientology, astronomy and astrology.

Space advocates and enthusiasts should understand that reason—which allowed us to go to the Moon—should guide every aspect of our lives. Societies and cultures will allow human begins to achieve the best within them to the extent that reason rather than mindless faith is their credo. How much more so will a creed of reason and the virtue and habits of rationality be needed on the Moon?

Responsibility—Each individual deserves freedom because we are each responsible for our own lives, beliefs, actions and moral character. Ayn Rand put it well when she said, "as man is a being of self-made wealth, so he is a being of self-made soul." The freedom offered in a new world, as will be available on the Moon, can unleash the creativity of every individual. Individuals should be free to experiment with their own lives. If they show us better ways, we all benefit. If they fail, they and we learn valuable lessons. But we each must appreciate that we are responsible for our own lives.

Justice—A basic principle of justice is that one respects the rights of every other individual to their own lives, actions, goals and dreams, that one not demand from others what is unearned. Social conflicts are the result of individuals or groups of individuals using force, usually government force, rather than voluntary exchange to extract goods, services or favored behavior from others. Unfortunately, the Apollo Moon landings, which were among the greatest achievements in human history, were made on money extorted from taxpayers.

Those settling the Moon can leave behind moral error of thinking that others owe them a living, and commit to the principal of voluntary exchanges and relations with others based on mutual consent.

Benevolence—While one might not owe any individual anything other than respect for their liberty, out of a society of individuals can emerge true benevolence. It is not an oxymoron to find benevolence in individualism. Men and women who value their own lives and who respect the lives of others will benefit by trading goods and services with others, and will be entertained, enlightened, educated and inspired by the plays, poetry, paintings, movies, music, scientific discoveries, engineering feats and every manner of human achievement of others. A society of individuals will be a society worth preserving. Thus individuals will want to help others where possible, not out of a joyless duty, but out of a cheerful desire to foster the kind of beauty a gardener does when tending to flowers, or a teacher does when cultivating young minds.

Those who wish to see the Moon settled are farsighted romantics, lovers of human potential and passionate pioneers who see new cities gleaming in the Sun on the Lunar surface. The ethos of reason, responsibility, justice and benevolence is most fitting for them, as it is for all human beings.

It is just as dangerous to expect utopia on the Moon as it is to expect it on the Earth. But the Moon offers an opportunity for a new and better civilization to emerge, just as America did in the past. A realistic but optimist understanding of the potential of free men and woman can create that better place for humanity in the heavens.

Edward Hudgins is executive director of the Objectivist Center and the Atlas Society, institutions that promote a culture of reason, individualism, human achievement and liberty. He is also editor of the Cato Institute book, *Space: The Free-Market Frontier*, and advocate for the Space Frontier Foundation and a founding board member of the Institute for Space Law and Policy. Hudgins, a policy expert in transportation, economic development and other matters, previously worked at the Heritage Foundation, the Joint Economic Committee of the U.S. Congress and the Cato Institute where he was director of regulatory studies and editor of *Regulation* magazine. He earned a Master's degree from American University and a Ph.D. from Catholic Univerity, both in Washington, D.C., in political philosophy. He has taught at universities in the United States and Germany.

In this, the last essay of this volume, science fiction author Allen Steele takes the long view, projecting his vision out a hundred years after the first return to the Moon. Steele has been compared to Robert Heinlein in his use of hard science and the types of gritty stories and characters with which he populates the future. This provocative piece lays out a key element of frontiers and their place in history that is often overlooked or ignored in cost based, strategic and academic discussions of the future. Call it the edge function—a place for those who don't fit in to go, to try out new ideas and social systems. Ironically, as distasteful as this or that social experiment might be to some, they represent the essence of what it means to be human, to grow, to evolve and explore that which is in us—frontiers even greater than the ones we find beyond ourselves.

Moon Age Daydreams

by Allen M. Steele

A year or so ago, a friend of mine e-mailed me an interesting news item: some guy in Romania was selling real estate on the Moon. According to a story carried by CNN.com, a gentleman in Bucharest was selling 177-acre lots on the Moon for $49.00. He began doing so after receiving authorization from Dennis Hope, the proprietor of the Nevada-based Lunar Embassy, who claims to have discovered a legal loophole in the 1967 U.N. Space Treaty which he says allows private ownership of land on the Moon. So far, most of these lots have been bought by Europeans who want to give their sweethearts something a little less ordinary than roses and chocolates for Valentine's Day.

Although it's touching to know that someone would give his girlfriend or wife a piece of the Moon, I have to wonder how well it would stand up to a serious legal challenge. I haven't reviewed the U.N. Space Treaty lately, but I suspect that, if any of these plots happen to be located in what might be considered prime lunar real estate, high-power attorneys employed by multinational space companies would go head-to-head against storefront lawyers representing folks who own deeds purchased for less than fifty bucks decades earlier.

More to the point ... if you can get past the giggle-factor inherent in the idea of someone selling land on the Moon, then you're on the way to accepting lunar colonization as something that may happen sometime later in this century. Not just small-scale base camps or even settlements of twenty or thirty people huddled together in modules buried beneath the regolith, but indeed settlements of hundreds—perhaps even thousands—of people living on the Moon.

For a few minutes, let's put away all the bar-graphs and pie-charts and schematic diagrams, and consider what it might be like to live there.

One of the most recent studies of near-future lunar colonization was done in 2000 by the Orbital Technologies Corporation (Orbitec) for the NASA Institute for Advanced Concepts (NIAC). The report, titled "Final Report on System Architecture Development for a Self-Sustaining Lunar Colony" represents a state-of-the-art portrait of a large, self-sustaining moon base.

One of the first terms which the Orbitec study seeks to define is what is meant by "self-sustaining," as opposed to "self-sufficient." A "self-sustaining" lunar settlement would be able to generate and recycle its own air, water, and food, making it capable of maintaining a large population for years at a time. However, it would still be dependent upon Earth for re-supply of complex tools, heavy machinery, electronics, and drugs. A wholly "self-sufficient" colony, on the other hand, would be completely autonomous, with the capability to manufacture everything it needs.

Let's assume that our lunar colony—let's call it Aldrinburg, after Buzz Aldrin, the second man to walk on the Moon—falls somewhere between these two levels. It's a very large settlement, with a full-time population of around 1,200 people, living in a medium-size impact crater about five miles in diameter. The crater has been covered with a flat, semi-rigid roof made of some airtight material held in place by cables anchored to the crater rim; the roof is covered with lunar regolith, about three to fifteen feet in depth, which provides radiation protection for the inhabitants. The interior crater walls have been carved out to provide apartments along circular terraces, much like the cliff dwellings of ancient Anasazi Indians. The enclosed crater floor is a vast open space, encompassing not only buildings but also farm fields, grazing areas for livestock, perhaps even small ponds for raising fish.

The business of Aldrinburg is business. A few visitors from Earth come up every month. There's a small bed-and-breakfast on the crater floor, complete with an outdoor cafe and its own tennis court. They tend to be either business travelers or relatives of those who live there. Tourists prefer Apollo, the luxury resort in Mare Tranquillitis a few hundred miles away. Aldrinburg (or "Buzztown," as it's affectionately known by the locals) is a company town; most people work in the nearby industrial park, a sprawling collection of domes along the roads leading from the crater to the landing field. The town's primary exports include helium-3 (refined from regolith for use in fusion reactors on Earth), lunar oxygen and hydrogen (sold as fuel for cis-lunar spacecraft), and silicon for computer chips and solar cells manufactured back home. Wheat, soybeans, and corn cultivated on its farms are sold to other lunar colonies; there's also another agricultural export, too, but we'll get to that in a moment.

The standard of living is fairly high. Most of Aldrinburg's residents are employed by Earth-based corporations that have signed them to two-and-three year contracts, while others belong to employee-owned enterprises. The majority of the inhabitants recently migrated here from various countries on Earth, yet a few are second-generation "Moondogs," whose fathers and mothers were among those who came to the Moon in the early part of the century. A small handful of these—children for the most part, the eldest in their teens—were actually born and raised on the Moon. You can tell who the Moondogs are because they tend to be tall, pale, and skinny, staying out of direct sunshine and lower gravity notwithstanding.

Buzztown is self-sustaining, but it is not self-sufficient. Nitrogen is a valuable commodity; it has to be imported from Earth by the tankful, both as breathable air and as a crop fertilizer. Water is derived from lunar oxygen and hydrogen, but chlorine and other purifiers also have to be brought up. Machine shops in Aldrinburg can manufacture small tools and factory equipment, but anything more complex than a hammer or a drill-bit has to be purchased "down there." Ditto for high-end electronic equipment; a new desktop computer can cost up to a quarter-million dollars to be shipped to the Moon—assuming that DNA-based quantum computers haven't been developed by this time—so old systems are continually upgraded rather than junked. Herbal medicines are plentiful, but vitamins, minerals, antibiotics, and even salt tablets are costly items. The local hospital hoards its blood supply as if it was gold bullion. Everyone who comes here even for a short stay has to be thoroughly vaccinated before leaving Earth.

As a result, paychecks tend to disappear fairly quickly. A four-hour shift driving a regolith combine will get you a bottle of aspirin. Two weeks planting cabbage will earn enough for a new chip for your PDA. Three years working eight hours a day in the lunox plant, and you might consider tearing up the lease on your three-room apartment in the crater wall and buying it for yourself. The money is good, sure, but the inflation rate would be steep.

Nonetheless, the people of Aldrinburg consider themselves pioneers. They'd have to be. Anyone who decides to take a job on the Moon, perhaps bringing along his or her family as well, better not get homesick easily. Most of the companies have a 90-day "wash-out" clause in their contracts. If a supervisor notices a new employee spending a lot of time gazing up at Earth, getting in quarrels over minor nuisances, or otherwise displaying signs of depression, then the company has the right to ask the new arrival to submit to personal counseling. If the psychiatrist judges that person to be unfit, then the company would decide whether to terminate his contract and send him home. The wash-out rate might be ten percent; every so often, someone comes up who just doesn't belong here.

Many of Buzztown's citizens moved here because they've left something behind. Some are getting over failed marriages. Some are adventurers restless with life back home. Some were too weird for wherever it was that they came from. Some are running from the law, taxes, parents, or other problems. And, some simply looked up at the Moon late one evening and decided that, all things considered, they'd rather be someplace else.

So now, here they are: 1,200 men, women, and children, living together in a place nearly a quarter of a million miles from the cradle of humanity. There is blistering heat for two weeks out of each month and freezing cold for the other two. The seas are basalt, the mountains look like melted lead, and everything is rationed, even the very air you breathe. Your home is a cave, your job is in a dome. Make a person-to-person phone call home for Mother's Day, and you'll notice a slight time-delay between what you say and what Mama tells you.

It's no wonder Aldrinburg's residents would tend to be independent-minded. Frontiers attract persons who aren't easily regimented, and considering the unique environment they're living in, people in Buzztown will be folks who are used to thinking outside the box. They might be employed by multinational corporations, but there isn't going to be a suit and tie among them ..or if there is, then he or she would keep that outfit hanging in the closet.

In fact, it wouldn't surprise me if lunar society closely resembled that of a small town in rural America in the early part of the 20th century, rather than the authoritarian, quasi-military model favored in most science fiction stories. Local government would involve democracy by consensus; important decisions would be made during town meetings presided over by a board of selectmen, with an elected mayor or a town manager wielding the gavel and every citizen given the right to voice his or her opinion. Children would all attend the same school, from first grade through high school. Or they might be collectively home-schooled, with parents assuming the role of tutors, each father or mother taking a turn every day to stay home to teach the kids while the others go off to work.

Co-ops would spring up. With everything in short supply, it would make a lot of sense for people to pool their spending money and purchase the goods they need from Earth in bulk quantities. This would also go for things colonists make themselves, whether it be food grown on the farms, new clothing, herbal medicines, hand-crafted items like kitchen utensils and furniture, and so forth. The co-ops would also serve as another social thread; seeing each other at the store, neighbors would gossip, make arrangements for dinner or babysitting, exchange recipes, and so on.

For recreation, one might expect to find gardening clubs, local theater ensembles, book discussion groups, and intimate little salons where like-minded individuals get together once a month for wine and intellectual discussion. The more active types would find other ways to amuse themselves. Geology clubs would form, in which amateur rock-hounds would venture out into the "lunar boonies" to collect interesting samples. Climbing groups would tackle mountains or volcanic craters, while spelunkers would explore lava tubes. There might even be a form of drag-racing, where souped-up rovers compete against each other on flatlands cleared of boulders.

One might also expect lunar society to evolve in ways that might seem immoral or illicit by present-day terms. For example, in my novella "The Weight," I had group-marriages taking place in a mid-21st century lunar colony, with triads of previously monogamous couples formally bonding together to create small clans in order to form lasting partnerships among close friends, and to raise their children as members of the same extended family. Such a thing would be considered polygamy today, and thus illegal in the United States, but in an isolated society like Aldrinburg it might be considered an attractive alternate lifestyle.

Another practice that would make sense in a lunar colony would be the cultivation of cannabis sativa—hemp by one name, marijuana by another. Until the passage of the Marijuana Stamp Act of 1933, cannabis was legally grown in the United States, and at one time in the 1800's it was the country's second-largest cash-crop. And for good reason: cannabis can been used to make everything from medicine to motor oil. The original copies of the Declaration of Independence were printed on hemp paper, and during World War II the lines of parachutes worn by American airmen were made of hemp rope. I regularly wash my hair with hemp shampoo, use hemp body lotion to keep my skin from going dry during winter, and until lately carried my cash and credit cards in a wallet made of hemp cloth ... and I'd pass the strictest urine test with flying colors, no pun intended. Cannabis sativa is a hardy weed that's easily grown under conditions that would kill cotton, and the psychoactive chemicals that make the leaves and buds of its female plants worth smoking by some can be rendered inactive through cross-pollination with male plants. So for a lunar colony like Aldrinburg, hemp cultivation as a major crop would be a reasonable alternative to having everyday items shipped up from Earth.

Without a doubt, such a society would be viewed by many on Earth as sybaritic, anarchistic, utterly immoral. Politicians would demand a crackdown. Churches would dispatch Bible-wielding missionaries to convert the God-less heathens. Yet, no government on Earth would have much control over a colony so far from home, and missionaries might find few converts among those who already consider themselves to be upstanding citizens.

In the meantime, a line gathers at the spaceport. Rebels and runaways, jack-mormons and bohemians, artists and holy fools—everyone who has ever wanted to leave home, whether it be by covered wagon, steamship, airliner or space shuttle—wants to light out for the frontier.

The Moon would become the New World.

All these things—government by consensus, home-schooling for children, co-op stores, salons and hiking clubs, group marriages, cannabis as a major crop—might sound as if a mid-21st century lunar colony would be little more than a late-20th century hippie commune. And so it might be. Skeptics are entitled to question these prognostications.

Likewise, it could be argued that lunar colonies would become tightly-controlled societies whose citizens are under constant surveillance, with even the slightest deviation from social mores punishable by arrest, incarceration, even expulsion. However, I have a hard time believing that such a regime would last very long. People just don't take crap like that for no reason, particularly if they're originally from free societies, and it would be only a matter of time before a little lunar Hitler and his cronies were overthrown.

Whatever form it might take, though, I have little doubt that daily life within a self-sustaining lunar colony would be much different than what we know today. It's foolish to assume that it would be much like living in a middle-American suburb or even an urban apartment building. New frontiers pose new challenges; the people who choose to live there find their own solutions. This is the genius of the human species: we adapt to strange environments, and find new ways of living unimagined by previous generations.

And this is the reason why, once you've landed on the outskirts of Aldrinburg and boarded the six-wheeled bus that transports you from the lunar shuttle to the subsurface entrance to the crater, you will see a strange thing on the side of the road: a twenty-foot replica of the Statute of Liberty. Welded together from bit and pieces of scrap metal, it was placed there several years ago by the men and women who built Buzztown.

They knew who was following them to the Moon. And why the best of them are going to stay.

Allen Mulherin Steele, Jr. became a full-time science fiction writer in 1988. He earned his B.A. in Communications from New England College and an M.A. in Journalism from the University of Missouri. His novels include Orbital Decay, Clarke County Space, Lunar Descent, Labyrinth of Night, The Jericho Iteration, The Tranquillity Alternative, A King of Infinite Space, Oceanspace, Chronospace, and Coyote. He has also published four collections of short fiction and his novellas have won numerous Hugo Awards. Steele serves on the Board of Advisors for both the Space Frontier Foundation and the Science Fiction and Fantasy Writers of America, and is a former member of the SFWA Board of Directors. In April, 2001, he testified before the U.S. House of Representatives, Subcommittee on Space and Aeronautics, in hearings on the future of space exploration. He lives in western Massachusetts with his wife Linda and their two dogs.

I drafted this in July 1999 at the Foundation's Return to the Moon Conference. The idea was to attempt to codify the concepts and precepts those of us who believe humanity should return to the Moon. Feel free to make your own copy, get some folks to sign it, and send it to your friendly local politician.

The Lunar Declaration

Recognizing it is the dawn of a new millennium, and the human spirit cries out for new beginnings;

Recognizing the need for new challenges to the human spirit, new domains for the exploration of human freedoms and the rapidly growing pressures on our biosphere and natural systems;

Recognizing humanity has developed the ability to routinely access and utilize the space near Earth and maintain human habitation in the space environment;

We the undersigned, citizens of Earth, do hereby declare that it is the duty and responsibility of our species to expand our civilization and the biosphere of our home world outwards into space.

We further declare that it is our duty to assure that this movement is safe, supportable, sustainable, and unstoppable.

First explored by human beings several decades ago and given its proximity to our home world, its location on the edge of the Near and Far Frontiers of human exploration, its bountiful resources, its ability to serve as a platform for further exploration and as a nearby location for our first human habitats on another planet, we believe the Moon represents the next and most vital step for humanity as we expand beyond Earth orbit.

Be it as a training base for future human explorers of Mars and other worlds, a supplier of precious materials for the development of clean energy on Earth and construction in the space between planets, a home to observatories that will probe the cosmos, a location for commercial enterprises including hotels, or simply as land to be settled and owned by individuals who are willing to stake their lives and fortunes to open its bounties; the Moon represents a new opportunity for an unprecedented partnership between the public and private sectors that will result in savings to taxpayers and profits to those willing to take the financial risks.

We believe there is an appropriate role for all in this endeavour, with private industry providing services and supporting operations and the government providing a regulatory environment and acting as a good customer as it fulfills its legitimate needs to develop the technologies and systems necessary to explore the far frontier

Therefore, we call on the people of Earth and its governments, industries and institutions, each acting in their appropriate roles, to join together in a renewed and united effort to seek these synergies, mold a new unified approach to opening the frontier, and create the financial, legal and policy incentives that will catalyze this effort.

To summarize: We the undersigned, citizens of Earth, do hereby declare it is time to Return to the Moon. This time to stay.

Moon Facts

The Moon, also known as Luna, orbits the Earth at an average distance of 238,906 miles (384,400 km). Luna's rate of rotation matches its orbital period of 27.32 days precisely, thus it appears to have one face locked toward the Earth. It is a geologically dead world, about 1/6 the diameter of its parent. This large relative size as compared to the Earth has led many astronomers to consider the Earth and Luna to be a binary system. (Most of the moons in our Solar System are relatively tiny compared to their parent planets.)

Luna has three basic types of terrain: bright highlands (once thought to be continents), dark basaltic lowlands (once thought to be oceans) and extensive cratering.

The Moon is more than just another distant planet; due to its proximity, it is a part of human culture. It affects us directly through our very biological makeup, with the pushing and pulling of the ocean tides. It also has become important to our society, acting as a market of time, a residence for gods and demons, and, since we decided to go there, as our first off-planet destination.

Luna is the only other body in the Solar System to have been visited by human beings, and is also the only other body for which we have actual soil samples.

History and Explanation

Luna and the Sun are the two most visible objects in the sky. For this reason, they both share a rich history in the myths and fables of humankind. Early civilizations endowed the Moon with Godhood, or more often than not, Goddesshood.

Visible only as a pale ghost during the day, and in its glory as ruler of the night, many people thought Luna represented dark forces, or that staring at the Moon would produce insanity (thus the term, "lunatic.")

Thankfully, some people did stare at the Moon, and over time these observers began to learn about our tiny sister world. Some of the earliest observations were made by fisherman and sailors. The rising and falling of the tides was related to Lunar cycles before recorded history began.

Even until 1609, when Galileo trained his telescope upon the Lunar surface, the Moon had been interpreted as a disk, or even a hold in the heavenly cloak of stars by many, although most believed it to be the closest of a set of perfect spheres circling the Earth.

What Galileo saw was a rugged terrain covered in craters, mountains and smoother areas that he thought were large seas ("maria," in Latin). Others of the time, using similar instruments, also believed Luna to be dotted with small seas or oceans. Thus today we hear the names "Sea of Tranquillity" or "Ocean of Storms" applied to the Moon's dark lowland plains.

In the next century, as Isaac Newton's ideas of planetary motions spread, observations of the Moon's orbit and perturbations became almost a field unto itself, with well-known astronomers and mathematicians, such as Giovanni Cassini, Joseph La Grange, and others, developing detailed theories and tables to test the new ideas against the Moon's relatively easy-to-observe motions.

In the 18th century, observations continued and more maps were drawn, as ever more powerful telescopes went on line. At one point, there was a scare that settlements were seen on some of the "continents," but the reports eventually faded away.

Still, it wasn't until the space age that a picture of the real Luna could be drawn. Until visitations by probes and humans in the 1960s, the Moon's highlands were thought to be covered in jagged tooth-like mountains, and the lowlands were filled with dust, possibly hundreds of feet thick.

Since 1959, we have virtually bombarded the Moon with probes, and following is a list of the high points. In 1959, the Soviet space probe Luna 2 crashed on the far side. A month later, this was followed by Luna 3, which was able to return our first images of the distant face. It revealed a much more cratered surface, and almost no smooth maria. The first American attempts at reaching the Lunar threshold were dismal failures. Over a dozen probes either failed or missed their targets by wide margins. (Interesting to note that it was exactly during these failures that President John F. Kennedy announced Apollo's goal of putting a human on the Moon less than ten years later.) In 1964 and 1965, Rangers 8 and 9 finally snapped excellent close-up images as they crash-landed on the surface.

During February 1966, the Soviet Luna 9 lander successfully touched down and sent back the first ever photographs from the surface of another body. Then for one year, between August 1965 and 1966, a series of American Lunar Orbiters circled and mapped the Moon to a resolution of 2 meters (6.5 feet). Simultaneously, the United States began landing Surveyor spacecraft on the surface, providing thousands of clear photos.

In 1968, the first humans flew around the Moon in Apollo 8. They were followed by three more astronauts in Apollo 10, which tested Lunar-landing techniques. Then came the big one, when on July 20, 1969, two human beings stepped upon Luna's Sea of Tranquillity, in one of the greatest moments in human history.

The Apollo series transported a dozen men to and from the Lunar surface, including one geologist, Harrison Schmidt, on Apollo 17. Their observations, and the information provided by the Lunar Orbiters, revolutionized our knowledge of the Moon. In total, Apollo returned over 382 kilograms (842 pounds) of samples to Earth, and hundreds of experiments were undertaken.

Little else was done regarding the Moon for almost 20 years—that is, until 1994, when the U.S. Ballistic Missile Defense Organization placed a test vehicle with high-resolution cameras in Lunar orbit. Named Clementine, after the miner's daughter, the tiny low cost probe returned some of the highest resolution images yet returned of the far side, and created the most detailed map we currently have of the Moon. The probe also appears to have detected trace amounts of long-frozen water in the Moon's permanently shaded polar craters, a discovery that would once again transform our image of the Moon, and would certainly dictate that any permanent future human outposts should probably be located nearby.

Geology

There are three major theories, which might explain the origin of Luna. The first and least likely is that both Moon and Earth were both formed out of the dust of the early Solar disk, and that the smaller amount of debris which formed the Moon simply accreted close enough to the Earth to become its satellite.

The second theory holds that Luna was actually created elsewhere in the solar system, and was later captured by the Earth's gravity. The difficulty with this idea is that the great speed of the un-captured Luna would need to have been dampened by a much stronger gravitational field than is possessed by Earth, in order to capture the wandering world and bring it into the orbit it now maintains.

Finally, there is the so-called "fission" theory. This idea supposes a newborn and rapidly rotating Earth (with a rotation rate of only a few hours). As the world spun, it flattened at the poles to such a degree that a large chunk of its outer crust broke free. This theory explains much, but fails to address the loss of momentum needed to slow the two bodies down to today's speeds.

Unlike the dynamic Earth, Luna has been geologically dead for around three billion years, and has no atmosphere to shape its surface. Any changes to its features have been made by outside forces, mainly the impact of meteors.

There are several types of geological features evident on the Lunar surface. The two main divisions are the bright highlands, composed of more silica-based materials, and the darker lowlands, which are made up of more metal-laden materials. Cratering is extensive, especially on the far side, where the giant flows of basaltic lava, called maria, have not obscured them. The maria themselves, which dominate the near side, are much more recent in origin than many of the craters which have been filled by their flows.

The internal structure of Luna is fairly simple. There is a crust averaging about 70 km thick (it appears to be thicker on the far side), which is itself layered over by the volcanic basalt flows of the maria. Under this zone lies the mantle, which is geologically inactive, except for the extremely mild and occasional "Moonquake," due in large part to the tidal interactions between the Earth and the Moon. There may or may not be a core at the center of the Moon, although the almost absent magnetic field indicates that if it does exist, it is quite small.

Major tourist attraction

From Luna, one can get quite a good view of Earth. It also may well possess large lava tube caverns, formed at the same time as the maria. The one-sixth Earth-normal gravity field is also an attraction for the elderly and disabled, who could discover new physical freedoms. Keep in mind, we do not yet know whether permanent residents, or those born on the Moon, would ever be able to return to Earth.

Surface Conditions

The Moon has not atmosphere to speak of. The vacuum of space literally reaches its surface. Without a moderating atmosphere, the temperature can reach a high of 105 degrees Celsius (221 degrees Fahrenheit) at midday, to –155 degrees Celsius (-247 degrees Fahrenheit) during the Lunar night. The Moon does have dust, and it was recently discovered that as the day/night terminator passes over its surface, this dust rises some height before settling back down. At times, this effect produces a slight coloring above those areas affected, as sunlight is bounced and refracted by the dust particles.

Valuable Resources

Someone once said that if a higher power wanted to give the people of Earth a gas station, it would have given us a Moon. Well, look outside your window tonight. There it is!

The Moon is rich in several materials useful for human exploration and settlement. Generally speaking, Lunar soil is several times richer than Earth soil, in titanium, aluminum, and calcium. It has less iron available, but there are usable quantities, and much of this iron is un-oxidized and can be retrieved using magnets.

Throughout the Apollo Luna samples, there is oxygen bound into several mineral compounds, sometimes averaging over 40%. This oxygen can be extracted, using several methods, many centered on the use of solar energy, abundant on the Lunar surface. Considering the obvious need for oxygen by terrestrial life, and that standard rocket propellant is over 70% oxygen by mass, there are very good arguments for creating Lunar oxygen refineries.

For example, a spacecraft could fly from the Earth to Luna, carrying only enough propellant to safely land, and a relatively small store of hydrogen on board, then refuel at the Lunar filling station and return to Earth. (Actually, even in the trace amounts which exist on the Lunar surface, some native hydrogen would be available, as it could be collected from the large amounts of that soil which would be processed anyway (for oxygen and the other uses mentioned below.)

Several experiments have also been conducted on the making of glass and fiberglass composites, using Lunar soil. Princeton's Space Studies Institute has been the leader in this field, and Japan's Shimizu Corporation has funded research into the manufacturing of waterless Lunar cement.

Recent imagery from the U.S. military's Clementine orbiter also indicate that there may be long-frozen deposits of ice hidden under the shadows of some polar craters. If so, the problem of the Moon's lack of hydrogen becomes much less of an obstacle to development.

Finally, there is Helium 3, an isotope of helium, which, according to physicists, can be used to generate fusion reactions, which are cleaner than those which could be created using current methods. Helium 3 is a byproduct of the Solar fusion reaction, and is believed to be deposited in a very fine layer across the Lunar surface by the Solar Wind. Since fusion has yet to be achieved, and very little of this material goes a very long way, one mission to collect a few kilos of the stuff would suffice for all of our Earthly needs for decades yet to come.

Future Exploration

As there has been so long a gap since our first forays proved it was within human reach, scientists, engineers, entrepreneurs and dreamers have created a massive amount of planning for exactly what to do with, and on, the Moon. The ideas range from using the Lunar surface as a repository for the complete works of humanity, to be stored in case the Earth is destroyed or do ourselves in, to propellant factories, helium 3 mining missions, and the construction of large Lunar far side astronomical observatories, shielded from the light and noise of the Earth.

There are ideas for tourist hotels, spas, resorts and adventure sports arenas. The largest concepts involve the use of Lunar soil, to build large solar arrays with which to supply clean energy to the Earth. These would either be constructed on its surface (as suggested by Dr. David Criswell) or in free space (as suggested by Dr. Peter Glaser).

(Discuss Clementine and Lunar Prospector)

Currently however, there are no funded government missions to return. Many n the Russian and U.S. space agencies have a "been there, done that" attitude about Luna, which is appropriate for explorers, but not if one plans to actually use the Moon as a lever for human expansion into space. In that case, Lunar research would now be moving into high gear, as we tried to learn how to use it for the good of humanity. As space settlement and development is not now the goal of any major space program, we cannot expect the governments of the world to invest in such goals.

But, hold it; this may not be so bad. Once again, we have an opportunity for the private sector to follow the trail blazed by the government.

You see, Luna is unique among the planetary bodies in the Solar System. It is close enough to the Earth, and requires a small amount of thrust to escape its gravity that commercial ventures there would at least be feasible. We also have the benefit of several government missions' worth of data, and research on its makeup and conditions, from which to draw information about its resources and how they might be utilized.

Another interesting feature of the Moon is that the time needed for a radio signal to be sent and returned from its surface is low enough that robots and tele-operated vehicles can be run by operators on Earth.

No more government employees should be sent to the Moon for artificial reasons of prestige or loosely disguised science. It is time to Return To The Moon. But this time it will be to develop the resources it offers us as we bootstrap our way into space.

MOON STATISTICS

Moon/Earth Comparison Bulk parameters

	Moon	Earth	Ratio (Moon/Earth)
Mass (1024 kg)	0.07349	5.9736	0.0123
Volume (1010 km3)	2.1958	108.321	0.0203
Equatorial radius (km)	1738.1	6378.1	0.2725
Polar radius (km)	1736.0	6356.8	0.2731
Mean density (kg/m3)	3350.00	5515.00	0.607
Surface gravity (m/s2)	1.62	9.80	0.165
Escape velocity (km/s)	2.38	11.2	0.213
Solar irradiance (W/m2)	1367.6	1367.6	1.000
Topographic range (km)	16.00	20.00	0.800

Orbital parameters (for orbit about the Earth)

Perigee (106 km)	0.3633
Apogee (106 km)	0.4055
Revolution period (days)	27.3217
Synodic period (days)	29.53
Mean orbital velocity (km/s)	1.023
Inclination to ecliptic (deg)	5.145
Inclination to equator (deg)	18.28 - 28.58
Orbit eccentricity	0.0549

Mean values at opposition from Earth

Distance from Earth (km)	384,467
Apparent diameter (seconds of arc)	1864.2
Apparent visual magnitude	-12.74

Distance From Earth:	225,745 miles
Length of a Day:	27.3 days
Radius:	1,080 miles
Weight:	81 Quintillion Tons
Surface Temp (Day):	273° F
Surface Temp (Night):	- 244° F
Gravity At Surface:	0.1667 g (1/6 Earth's)
Orbital Speed	2,287 mph
Mean orbital velocity (km/s)	1.023
Driving time by car (@70 mph):	135 days
Flying time by rocket:	60 to 70 hrs.
No. of Men Who Have Walked on Surface:	12
Age of Oldest Rock Collected:	4.5 Billions yrs.
Rocks Collected By Apollo:	842 pounds
Widest Craters:	140 miles (dia.)
Deepest Craters:	15,000+ (ft.)
Highest Mountains:	16,000+ (ft.)

Comparitive

	Earth	Moon
Equatorial diameter	12,756 km	3,476 km (2,160 miles)
Surface area	510 million square km	37.8 million square km
Mass	5.98 x 10E24 kg	7.35 x 10E22 kg

Crust

Earth Silicate rocks. Continents dominated by granites. Ocean crust dominated by basalt.

Moon Silicate rocks. Highlands dominated by feldspar-rich rocks and maria by basalt. Mantle Silicate rocks dominated by minerals containing iron and magnesium. Similar to Earth.

Core Earth Iron, nickel metal

Moon Same, but core is much smaller.

Atmosphere (main constituents)

Earth 78 % nitrogen,21 % oxygen

Moon Basically none. Some carbon gases, but very little of them. Pressure is about one-trillionth of Earth's atmospheric pressure.

Temperature Earth Air temperature ranges from -88°C (winter in polar regions) to 58°C (summer in tropical regions).

Moon Surface temperature ranges from -193°C (night in polar regions) to 111°C (day in equatorial regions).

Apogee Books Space Series

#	Title	ISBN	Bonus	US$	UK£	CN$	
1	Apollo 8	1-896522-66-1	CDROM	$18.95	£13.95	$25.95	_____
2	Apollo 9	1-896522-51-3	CDROM	$16.95	£12.95	$22.95	_____
3	Friendship 7	1-896522-60-2	CDROM	$18.95	£13.95	$25.95	_____
4	Apollo 10	1-896522-52-1	CDROM	$18.95	£13.95	$25.95	_____
5	Apollo 11 Vol 1	1-896522-53-X	CDROM	$18.95	£13.95	$25.95	_____
6	Apollo 11 Vol 2	1-896522-49-1	CDROM	$15.95	£10.95	$20.95	_____
7	Apollo 12	1-896522-54-8	CDROM	$18.95	£13.95	$25.95	_____
8	Gemini 6	1-896522-61-0	CDROM	$18.95	£13.95	$25.95	_____
9	Apollo 13	1-896522-55-6	CDROM	$18.95	£13.95	$25.95	_____
10	Mars	1-896522-62-9	CDROM	$23.95	£18.95	$31.95	_____
11	Apollo 7	1-896522-64-5	CDROM	$18.95	£13.95	$25.95	_____
12	High Frontier	1-896522-67-X	CDROM	$21.95	£17.95	$28.95	_____
13	X-15	1-896522-65-3	CDROM	$23.95	£18.95	$31.95	_____
14	Apollo 14	1-896522-56-4	CDROM	$18.95	£15.95	$25.95	_____
15	Freedom 7	1-896522-80-7	CDROM	$18.95	£15.95	$25.95	_____
16	Space Shuttle STS 1-5	1-896522-69-6	CDROM	$23.95	£18.95	$31.95	_____
17	Rocket Corp. Energia	1-896522-81-5		$21.95	£16.95	$28.95	_____
18	Apollo 15 - Vol 1	1-896522-57-2	CDROM	$19.95	£15.95	$27.95	_____
19	Arrows To The Moon	1-896522-83-1		$21.95	£17.95	$28.95	_____
20	The Unbroken Chain	1-896522-84-X	CDROM	$29.95	£24.95	$39.95	_____
21	Gemini 7	1-896522-82-3	CDROM	$19.95	£15.95	$26.95	_____
22	Apollo 11 Vol 3	1-896522-85-8	DVD*	$27.95	£19.95	$37.95	_____
23	Apollo 16 Vol 1	1-896522-58-0	CDROM	$19.95	£15.95	$27.95	_____
24	Creating Space	1-896522-86-6		$30.95	£24.95	$39.95	_____
25	Women Astronauts	1-896522-87-4	CDROM	$23.95	£18.95	$31.95	_____
26	On To Mars	1-896522-90-4	CDROM	$21.95	£16.95	$29.95	_____
27	Conquest of Space	1-896522-92-0		$23.95	£19.95	$32.95	_____
28	Lost Spacecraft	1-896522-88-2		$30.95	£24.95	$39.95	_____
29	Apollo 17 Vol 1	1-896522-59-9	CDROM	$19.95	£15.95	$27.95	_____
30	Virtual Apollo	1-896522-94-7		$19.95	£14.95	$26.95	_____
31	Apollo EECOM	1-896522-96-3		$29.95	£23.95	$37.95	_____
32	Visions of Future Space	1-896522-93-9	CDROM	$27.95	£21.95	$35.95	_____
33	Space Trivia	1-896522-98-X		$19.95	£14.95	$26.95	_____
34	Interstellar Spacecraft	1-896522-99-8		$24.95	£18.95	$30.95	_____
35	Dyna-Soar	1-896522-95-5	DVD*	$32.95	£23.95	$42.95	_____
36	The Rocket Team	1-894959-00-0	DVD*	$34.95	£24.95	$44.95	_____
37	Sigma 7	1-894959-01-9	CDROM	$19.95	£15.95	$27.95	_____
38	Women Of Space	1-894959-03-5	CDROM	$22.95	£17.95	$30.95	_____
39	Columbia Accident Rpt	1-894959-06-X	CDROM	$25.95	£19.95	$33.95	_____
40	Gemini 12	1-894959-04-3	CDROM	$19.95	£15.95	$27.95	_____
41	The Simple Universe	1-894959-11-6		$21.95	£16.95	$29.95	_____
42	New Moon Rising	1-894959-12-4	DVD*	$33.95	£23.95	$44.95	_____
43	Moonrush	1-894959-10-8		$24.95	£17.95	$30.95	_____
44	Mars Volume 2	1-894959-05-1	DVD*	$28.95	£20.95	$38.95	_____
45	Rocket Science	1-894959-09-4		$20.95	£15.95	$28.95	_____
46	How NASA Learned	1-894959-07-8		$25.95	£18.95	$35.95	_____
47	Virtual LM	1-894959-14-0		$29.95	£22.95	$42.95	_____
48	Deep Space	1-894959-15-9	DVD*	$34.95	£22.95	$44.95	_____
49	Space Tourism	1-894959-08-6		$20.95	£15.95	$28.95	_____
50	Apollo 12 Vol 2	1-894959-16-7	DVD*	$24.95	£15.95	$31.95	_____
51	Atlas Ultimate Weapon	1-894959-18-3		$29.95	£16.95	$37.95	_____
52	Reflections	1-894959-22-1		$23.95	£16.95	$30.95	_____
53	Real Space Cowboys	1-894959-21-3		$29.95	£17.95	$36.95	_____
54	Saturn	1-894959-19-1	DVD*	$27.95	£18.95	$35.95	_____
55	On To Mars Vol 2	1-894959-30-2	CDROM	$22.95	£14.95	$29.95	_____
56	Getting off the Planet	1-894959-20-5		$TBA	£TBA	$TBA	_____
57	Return to the Moon	1-894959-32-9		$22.95	£15.95	$28.95	_____

* NTSC Region 0

Pocket Space Guides

#	Title	ISBN	US$	UK£	CN$	
1	Apollo 11 First Men	1-894959-27-2	$9.95	£6.95	$12.95	_____
2	Mars	1-894959-26-4	$9.95	£6.95	$12.95	_____
3	Launch Vehicles	1-894959-28-0	$9.95	£6.95	$12.95	_____
4	Deep Space	1-894959-29-9	$9.95	£6.95	$12.95	_____

Apogee Science Fiction

#	Title	ISBN	US$	UK£	CN$	
1	Edison's Conquest of Mars	0-9738203-0-6	$9.95	£6.95	$12.95	_____